Nathan Allen

Physical Development

The Laws Governing the Human System

Nathan Allen

Physical Development
The Laws Governing the Human System

ISBN/EAN: 9783337371524

Printed in Europe, USA, Canada, Australia, Japan

Cover: Foto ©berggeist007 / pixelio.de

More available books at **www.hansebooks.com**

PHYSICAL DEVELOPMENT

OR THE

Laws Governing the Human System

BY

NATHAN ALLEN M. D., LL. D.

MEMBER OF

THE AMERICAN MEDICAL ASSOCIATION, THE AMERICAN ACADEMY
OF MEDICINE, THE AMERICAN PUBLIC HEALTH ASSOCIATION,
THE MASSACHUSETTS MEDICAL SOCIETY.

BOSTON 1888
LEE AND SHEPARD Publishers
10 MILK STREET NEXT THE " OLD SOUTH MEETING HOUSE "
NEW YORK
CHARLES T. DILLINGHAM
718 AND 720 BROADWAY

PAGE 342.

INTRODUCTION.

———•———

PHYSICAL DEVELOPMENT is selected as a title for this work. Almost every article it contains has some reference to the development of the body, or the laws that govern the human system. For twenty-five years the writer has published a large number of articles in journals, magazines, and in pamphlet form. These papers have been of a practical character and had reference to some improvement or reform in the state of society. At the time of publication several of these papers attracted much attention, and were favorably noticed by the press, both in this country and in Great Britain. They have called forth a large correspondence with the writer, and a demand for several of these papers, from a distance, which still continues.

Some of these papers are out of print, and others are so scattered, or lost sight of, that they can not be found. Questions were discussed in these papers which were somewhat new, and upon which there were differences of opinion, and still other questions which were in advance of the times. It seems desirable that in some form these papers should be reproduced. Besides, the writer maintains that there are new and important principles in physiology, which were brought before the public, that should be more thoroughly discussed, as well as their various applications pointed out. It is only in this way that the truth can be ascertained, and new principles become settled. In justice to the writer, and with a view to the public benefit, good judges familiar with these papers advise that they be reproduced, in part, or in substance, so as to be made available. Instead of making a compilation or collection of papers, the writer has virtually made a new book, by writing several new papers, by re-writing others, and introducing only the better part of a few long articles.

In submitting this book to the public, two explanations are deemed necessary. *First:* The frequent use of, and reference to, physiology. This science is not only important and comprehensive, but is a modern science; is comparatively in its infancy, in respect to many of its applications. No estimate can be made of the great advantages in the way of education and human welfare that are yet to be derived from this science. *Second:* In stating and explaining certain views in the different papers, there is some repetition which, for a correct and full understanding of the subject could not well be avoided. As these principles were new and very important this course seemed justifiable.

Contents.

APPENDIX.

Biographical Sketch.

[Taken from the *New England Medical Journal* (monthly), February, 1884.]

NATHAN ALLEN, M. D., was born in Princeton,
Mass., April 25, 1813. His parents, Moses and Me-
hitable (Oliver) Allen, were both born in Barre, in the
same state. The Allen patronymic is borne by numerous
families in the Old and the New world; but that from
which Doctor Allen is a lineal descendant, was Walter
Allen, one of the original proprietors of Old Newbury,
and who died in Charlestown, Mass., in 1673.

The early years of Doctor Allen were spent on the
parental farm. Here he was habituated to all kinds of
exercise in farm-work, and consequently received the best
possible preparation in health and constitution for future
activities. At the age of seventeen he commenced aca-
demical studies, and, matriculating at Amherst college in
1832, he graduated in 1836. Having decided to enter the
medical profession, he determined to avail himself of the
best advantages which the country afforded. Accordingly
he repaired to Philadelphia, where he spent four years in
the study of medicine and attendance upon lectures.

During his residence in this city he was employed a
part of his time in proof-reading, and in the preparation
of material, connected with medical and physiological
works, being published at the celebrated press of Adam
Waldie. While engaged here he was brought into personal contact and correspondence with many distinguished
individuals. Among these were Dr. Charles Cadwell, of

Kentucky, the profoundest physiologist of his day ; the Hon. Horace Mann, of Massachusetts, who as an educator has never been excelled ; and also with George Combe, Esq., of Edinburgh (then on a visit to Philadelphia), unequaled as a practical philosopher.

Here, while a medical student, Doctor Allen learned something respecting the use of the pen and the power of the press, as well as the importance of every person *thinking for himself.* In the spring of 1841 he graduated "M. D.," presenting as his thesis, the "Connection of Mental Philosophy with Medicine." This essay was published in pamphlet form, and, while it attracted much attention at the time, it indicated the department of scientific investigation in which he has since become distinguished. In the autumn of 1841 Doctor Allen settled in Lowell, Mass., where he has since resided. Soon after commencing medical practice his attention was arrested by the great difference in the birth-rate between the native New England women, and the English, the Irish, the Scotch, the Canadian French, and the German ; and also by the small number of children in a New England family, compared with what it was fifty or a hundred years before. The size of the family, including all married couples, is now on an average only about one-third as large as it once was ; and the birth-rate of the Irish, English, Scotch, and German is, on an average, nearly twice as large as that of the strictly native New Englander. From many years of study and observation he became convinced that the "arts of prevention and destruction" were not sufficient in all cases to account for this great difference in birth-rate, but that there must be some other primary cause — that there might exist some difference or change in the organization itself to account for it.

This inquiry led to a wide range of studies, such as census and registration reports, works on population, vital statistics, and obstetrics. It also led to a careful observation in medical practice, of the differences in the physical development of women, and to the peculiarities in the physiology of different races and nations. As a result of these inquiries and reflections, covering a period of over twenty years, Doctor Allen became thoroughly convinced that Nature, or the Creator of all things, has established a *great general law of propagation applicable to all organic life.* He maintains that this is virtually a new discovery, and that it involves a most important principle in psychology and physiology.

As applied to human beings, it may be defined as follows: This law is based upon a normal or a perfect standard of the human system, where every organ of the body is complete in structure and performs fully all its natural functions. This principle implies that the body is symmetrical and well developed in all its parts, so that each organ acts in harmony with all the others. According to this principle, the nearer the organization approaches that standard, and the laws of propagation are strictly observed, the greater will be the number of children, and the better will be their organization for securing the great objects of life. On the other hand, if the organization is carried to an extreme development in either direction, viz., a predominance of nerve tissue, or of a low animal nature, the tendency in such families or races is gradually to decrease, and ultimately to become extinct. Thus people enjoying the very highest civilization, or living in the lowest savage state, do not multiply rapidly.

This law of population is strictly the normal standard of physiology, upon which other important principles are based. It is the standard of perfect health. In deviations from it are found weaknesses, diseases, and the abnormal classes, such as the blind, the deaf-and-dumb, the feeble-minded, the insane, etc. This principle furnishes the true law of longevity. The more perfect the organization is, and the more harmoniously all the functions of the body are performed, in accordance with the laws of health, the longer will be human life.

By means of this law of propagation, the mooted questions connected with the intermarriage of relations are easily explained. It also furnishes in physiology a standard upon which the laws of inheritance have their basis and solution. Without such a standard or general principle all these laws are enigmas. This law of physiology furnishes also the standard or model of beauty of the human form, so much sought after and admired by the Greeks, as well as the most refined and cultivated people of all nations. This law furnishes the key to the doctrine discussed by some writers, — "the survival of the fittest," — that among all peoples and nations there is a class which overcomes all others. This is nothing more or less than the doctrine of "natural selection," which in a few years has obtained so great notoriety.

But the investigations of Doctor Allen have not been confined exclusively to this law of population, but have taken in a wide range of topics, such as physical culture and degeneracy, insanity and state medicine, heredity and hygiene, education and temperance, divorce and the family institution, etc. etc. All the publications of Doctor Allen number over thirty distinct papers or pamphlets, and if brought together would make two large octavo volumes. Some of these papers are elaborate essays, and appeared in the leading quarterly journals and reviews ; other papers were read before learned societies or scientific associations.

Besides set papers carefully prepared, Doctor Allen has contributed largely to the local and religious press, and has written numerous reports for public institutions as well as charitable associations. His writings have been widely quoted by the press both in this country and in Europe ; and several of his papers have been republished entire in Great Britain. Some of them have been favorably noticed, not only by the leading medical journals in this country, but by such journals abroad as the *Medico-Chirurgical Review*, Dr. Forbes Winslow's *Psychological Journal*, the London *Medical Times and Gazette*, the London *Lancet*, and the Glasgow *Medical Journal*.

Doctor Allen has held important positions of trust and responsibility. In 1856 he was chosen by the legislature a trustee of Amherst college, and took a leading part in establishing in that institution the department of physical culture, which is accomplishing a grand work for the health of its students. Dr. Edward Hitchcock, professor of hygiene and physical education in the college for twenty-five years, in a printed report pays Doctor Allen the following compliment: "The title of this department was proposed by Dr. Nathan Allen, of Lowell, one of the trustees of the college — and of her graduates — of which he has been an early and long-tried friend, and a most devoted and faithful guardian of this department, of which he may well be styled the godfather."

In 1864 Doctor Allen was appointed by Gov. John A. Andrew a member of the Massachusetts State Board of Charities, and by re-appointments served throughout the entire existence (fifteen years) of the board. Being chairman part of the time, he contributed more or less to its annual reports, which have a standard value. The following extract from its fourteenth report is of special and permanent interest, and commends itself to the careful thought of the clergy, the political economist, the medical profession, the legislator, and all other classes whose thought and action have influence upon the body politic. He said : "No fact is better established in science than that there is a most intimate mental, as well as physical, relation between the parent and the child — between each generation and the succeeding one. This relation has been well expressed in the proverbs, — 'What is bred in the bone can not be whipt out of the flesh,' and 'Like begets like.' The hereditary relation has, we believe, a far greater agency in producing social evils and vices than has generally been conceived. This relation extends, by transmission, not only to the form of the body and the features of the countenance, but to every part of the system, — to the quality of the blood, — especially to those vital organs which give stamina of constitution and beget mental predispositions. Whatever agencies, therefore, are calculated to injure the body, or deprave the mind; to incapacitate an individual for self-support, or make him a corrupter of others, should certainly be exposed by the guardians of public charity."

In 1872 Doctor Allen visited Europe. He went as a delegate, commissioned by Governor Washburn, to the International Congress, which met in London to consider the matter of reform in prisons and other correctional institutions. Doctor Allen is connected with numerous medical and scientific associations, and occupies public positions of trust and honor in the place of his residence which we pass by unnoticed.

Doctor Allen's distinction arises mainly from his original researches into the laws and changes of population. His fame for the present and the future must rest chiefly, —*first*, upon his investigations into the changes of population in New England, — that this originates primarily from a change in physical organization, — and *secondly*, on the establishment of this fact, — the principle is not *local*, but there must exist a great universal law of human increase, extending to all peoples and races throughout the world.

When he first published some statistics showing the

decrease of the birth-rate in New England, compared
with what it once was, and to contrast this with the increased birth-rate among the foreign population, these
statistics surprised the public, and some were disposed to
throw ridicule upon them. But further statistics and
sound reasoning soon converted the ridicule into expressed
alarm, and directed attention to the *causes* for this decline
in the birth-rate of the old Puritan families.

Doctor Allen attributes it to various sources, such as
educational pressure, fondness for *mental* rather than
physical labor, too high a standard of living, etc. ; producing an undue development of brain and nerve-tissue
compared with other parts of the body. If a change of
organization affects increase of numbers ; if the tissues of
the human system, when carried to an extreme development in any one direction, tend to extinction, or if there
is a normal standard of organization better than all others
for propagation, and upon which a general law of increase
is based, it involves fundamental principles in physiology
of vast importance. It may for the present attract little
attention and have a slow growth, but, if true, it must in
process of time be universally recognized and have a powerful influence. Language can not express the value and
advantage of such a law. Inasmuch as it holds in its
destiny the highest welfare of the human race, its benefits
can not be inferior to those derived from the laws of
gravitation and electricity, which have to do mainly with
material things. Investigation and discussion will in time
determine here, as elsewhere, *what is truth*. And though
the public may at present look with distrust on this new
doctrine, and some may regard its advocate as an "alarm-
ist," yet if proved to be a fundamental law in physiology,
history will do him full justice.

The writings of Doctor Allen have already obtained
quite a notoriety. His essays on population have for years

attracted attention in France and Germany, and have often been quoted in Great Britain. And there are few men in the medical profession in this country whose writings have been quoted more generally by the press, or been referred to as authority in works having a large circulation. Not long since his *alma mater* honored him with LL. D., for his public services and in recognition of his original researches into the history and changes of population in New England. If the doctrines he has enunciated, in reference to physical development and human increase, shall prove true, higher honors are sure to come, and, in time, a far greater reputation awaits him.

Physical Culture in Amherst College.

A MOST successful experiment of combining physical exercises with mental culture has been made at this institution. So important is this experiment deemed in all its bearings that a full sketch of it is here given.

It is twenty-eight years since gymnastic exercises were first introduced into Amherst college as a distinct department of education. It was an important event, not only in this institution, but in the history of educational matters. Certain principles were then discussed and adopted which have had great influence in making the experiment there a success. These principles are fundamental, and should be kept constantly in view in all attempts to improve physical organization connected with education.

At the annual meeting of the trustees of Amherst college in 1860, the writer was appointed chairman of a committee to consider and recommend a plan and regulations for the government of this new department; and having served every year since on the "gymnasium committee," he is quite familiar with the origin and history of this movement. The questions successfully settled here will apply to every similar institution.

One of the first questions which confronted us in this enterprise was that the trustees and faculty of a college had no right, in introducing gymnastic exercises, to make them *compulsory*, — that all students must engage in them. It was said that all such exercises elsewhere had always been and were *voluntary*, and not made a necessary part of the curriculum of an institution ; that students did not come to college to have their bodies trained, but to edu-. cate their minds.   The answer was, if this education could not be prosecuted so successfully, nor the highest standard of scholarship be reached, without proper exercise of the body and the possession of good health, such exercise should certainly be required.

As the trustees made the laws of the college, and were for the time being the guardians of the students, they must know better than these pupils or their parents what kinds of exercise were best adapted for their highest welfare and improvement.   Lessons in mathematics and the languages are made compulsory, and if it is found that a certain training of the body enables the student to do this work easier and better, and by the same course he would maintain good health through college, this training of the body should by all means be commanded.

The second question was, in order to make these exercises successful and permanent, we must give them *character.*   It was understood that previously gymnasiums, manual-labor schools, and attempts at physical education generally, had failed, but there were good reasons for it.   One was, the *manner* in which it had been carried on was not adapted to develop and train the whole body in accordance with physiological laws.   Another reason was, that not sufficient importance had been attached to this kind of education by trustees and managers of institutions, by teachers, by the press, and public opinion generally.

The movers in this new department at Amherst determined to organize and start it in a manner to show that they regarded it of the very highest importance — not inferior to any other in college.   The first step was to place at its head a thoroughly educated physician, who should be a member of the faculty — equal in standing to any other teacher or professor in the institution. He should have the whole charge, not only of the gymnasium and its exercises, but he should be a teacher of anatomy and physiology, of hygiene and physical culture. Besides, he should have a general oversight of the health of students ; should have a watchful care of them at all hours, and caution them against overwork in study, as well as all irregular habits. In case a student was feeling unwell, or complained of sickness of any kind, he could freely consult this teacher as though a family physician. Thus by having a living teacher at the head, who is a member of the faculty and has charge of the health of students, it was intended to give this department the same position and prominence as any other branch of study. Still further: in making up the merit-roll of every student, of his rank in the class, this branch was to come into the account — his attendance, his deportment, his interest, and the improvement from the exercises as far as it could be ascertained. In case a student had an organic difficulty of the heart or lungs, or any other physical weakness that disabled him from going safely through with all the exercises required, he would be readily excused by the professor. No one could judge of this so well as a teacher of physiology. At the same time, if a student in other departments of the college wanted to get rid of any regular exercise, or get leave of absence, claiming to have some infirmity, to be sick, this medical teacher could easily settle the matter.

KINDS OF EXERCISE.

The third question to be settled was what series of exercises would be adapted to produce the best results in a college-course of study. The immediate object was to exercise all parts of the body systematically, and in such a variety of ways that the student should maintain

uniformly good health, and the whole system — including
the brain — be brought into the best possible working
condition.   Physiology, as well as experience, teaches that
what are called "light gymnastics" are best adapted for
this purpose.

In settling the kind of exercise, this depends upon what
you want to accomplish.   If you want to make expert ball-
players, or boat-rowers, or train the body to excel in other
out-door sports and games, particular muscles or parts of
the body must be exercised for this express purpose.   But
in an institution made up of large numbers, confined in
close quarters, all engaged in hard study, and wanting to
make the most of their time, it is found that light gymnastic exercises, accompanied with music, and systematically practised a half hour or so every day, work best.
At the same time, connected with them, other exercises,
such as marching, running, or singing of a sportive character, should be practised at times for amusement and
recreation.

There is still another class demanding special personal attention.   Suppose students come to college with a physical
system not well balanced — and there are many of this
character — some one part or organ weak and liable to
disease.   This, by careful examination, can be easily de-
tected.   As in a college, or any regular course of study,
great and continuous strain must be made upon the body,
it is highly important that this weak part should be known
and strengthened.   The highest measure of health is
where the whole physical system is well balanced — where
all the organs are perfect, or nearly so, in structure, and
each performs its own legitimate function.   This is the
highest or normal standard of health.

Now, by *special* physical exercise much can be done to
change and improve distinct parts of the body in this
direction.   But it will be seen at once that such special

personal exercises can not be reduced to a system for all indiscriminately to practise. This is individual work and must be a specialty in physical training. At the same time, light gymnastic exercises are calculated to help these personal weaknesses, or this defective organization, by improving the general health. These two systems of physical culture are both good in their place.

The question may be asked, What relation do the regular gymnastic exercises hold to out-door sports? We answer, as auxiliaries, as helpers, but not as substitutes; the objects are very different. Gymnastics are intended to exercise all the muscles in the body, and to improve the general health and strength ; while ball-playing, boat-rowing, and out-door sports call into exercise chiefly particular muscles and movements. Gymnastic exercises are calculated to promote a harmony, a balance of action and strength throughout every part of the body, while these sports increase the size and strength of certain portions of the body disproportionately to other parts.

Each class has its own specific benefits. There can be no question, however, but the interest and zeal in carrying on physical exercises of any one kind tend to increase it in all others ; especially where there is no competition. In gymnastics there is less danger of injury to the body, and far less risk to good morals. If these out-door sports are properly conducted, — not carried to extreme, — they may prove beneficial to students of all classes ; but at the same time, there is great danger of their abuse. After all, light gymnastics are altogether the best physical exercise for students in literary institutions. The danger is of carrying out-door sports and games too far, of consuming upon them too much time, of diverting attention from study, and creating an unpleasant competition.

That Amherst college has taken the lead in physical training and instruction, in respect to the laws of health, there can be no question; and that, also, great benefits have been derived from this course. Before presenting some facts bearing on this point, we give the testimony of an individual who ought to be a competent judge. Says President Eliot, of Harvard university : "It is to Amherst college that the colleges of the country are indebted for a demonstration of the proper mode of organizing the department of physical culture."

It can, we believe, be safely stated no that other large literary institution in this country, or in Europe, has for a quarter of a century conducted physical education so successfully and so thoroughly as this college. One of the secrets of this success has been that the department, at its very start, was placed upon high ground ; was treated with an importance and character equal to the classics, or mathematics ; and like these, its exercises were made obligatory, and its results, like these also, entered into the merit-roll of every student. But a stronger argument still was, that the students themselves became, from year to year, so convinced of the great advantages of these physical exercises in improving their health and perfecting their scholarship, that they would not give them up on any account.

While the present rank of scholarship can not be tested with what it would have been without these exercises, nor can it be compared with what it once was before they were introduced, there can be no question but that the present scholarship of students is of a higher grade and character.

HEALTH.

In the matter of health, the facts are more obvious. A careful account has been kept every year of the sickness or loss of time from every kind of complaint of the students, and it has been found to be steadily diminishing ;

but, what is more striking, less and less in each class. The freshman class have the most, the sophomore not so much, the junior still less, and the senior the least of all. Thus, year by year, each class steadily improves in health, showing the immediate benefits of such exercises. This is the reverse of what occurred thirty, forty, and fifty years ago. No statistics of sickness or loss of time from illness were kept at that time, but we distinctly remember many cases of fever and other complaints, of students breaking down in health and leaving college.

Another marked feature, resulting from physical training, we believe, more than from any other cause, is a change in the countenance and physique of students. This applies particularly to students in the advanced classes and to those graduating. Perhaps no one thing affords stronger evidence of good health and a high state of vitality than the human countenance, when carefully scanned by an expert sanitarian. Now, from an experience of fifty years with the college, and in attendance upon many commencements, we can testify that there has been a marked improvement in the countenance and physique of students.

Formerly there were more or less students with pale, sallow countenances, sometimes too spare, with a haggard, care-worn look, and without much expression ; but such a specimen is now seldom seen. Their countenances indicate a freshness and glow of health, with a clear skin and lineaments distinct and expressive, animated with highly arterialized blood. The body is better and more evenly developed in all its parts, and when moving or standing, its position is erect. The limbs perform good service, with movements easy and graceful, but at the same time prompt and vigorous. The whole appearance of students, with the changes of countenances and movements of the limbs, indicate a high state of physical health, vigor, and strength.

There is still another advantage gained, the value of

which can not be estimated in figures, nor fully described in language. By means of gymnastics and instruction in hygiene, the constitution of the student has been greatly strengthened, and regular habits have been formed favorable to good health, not merely while in college, but that will last through life. The student has thus laid the foundation for good health in all his future years. At the same time he has accumulated an amount of knowledge in respect to the laws of health, which will become more and more valuable.

There are still other advantages from this gymnastic training. It is an essential aid in securing better discipline in the institution. These exercises serve to give a safe vent to an excess of animal spirits, which otherwise might result in acts of mischief or trouble of some kind. This physical training is calculated to develop not only all parts of the body, but to make it symmetrical and well-balanced throughout. Such an organization tends to give its possessor self-reliance and self-control, by means of which he can turn to better account the activities of both the body and the mind.

We have stated that light gymnastics afford the best kind of exercise for students. They harmonize with the laws which regulate the growth and changes in the various organs of the body ; they are convenient for use, and economize time ; they can be directed and controlled better than out-door sports and games. In fact, the great objection against intercollegiate sports and games is, they can not be controlled or regulated by any one united power; there are constant friction and complaint, and not unfrequently ill-feeling and bad temper.

EXAMPLE AND INFLUENCE OF PHYSICAL CULTURE.

There is another point that deserves notice : The example in starting physical culture at Amherst, and its influence.   It is reported on good authority that over fifty large institutions in our country have either adopted some regular system of physical culture, or are making preparations for the same.   So intimately connected is a proper care and development of the body with mental and moral improvement, that this reform can not go backward or remain stationary.   The more thoroughly the interdependent relations between the mind and body are understood, the greater will be the value attached to a sound, healthy, and well-trained body.

In the summer of 1861 Dr. Edward Hitchcock, then a teacher in Williston seminary, and a graduate of the medical school of Harvard university, was invited to take charge of this enterprise.   The remarkable success that has attended his labors and instructions here for twenty-five years affords the best evidence of his peculiar fitness and qualifications, that "the right man is in the right place."   When this department started, some looked at it as a doubtful experiment; others feared it would be an encumbrance upon the institution; but the general verdict now, we believe, is, the college could not well get along without it.

In this sketch of physical culture at Amherst some notice should be taken of the superior advantages there for carrying on this work.   The first gymnasium, erected in 1860, became too small and inconvenient as the classes grew larger.   As a result of the deep interest felt in this department by one of its own students, a graduate of the class of 1879 (Mr. C. M. Pratt, of Brooklyn, N. Y.), the college is chiefly indebted for its new, magnificent building, called "Pratt Gymnasium."   It was planned by Dr. Edward Hitchcock, after many years' experience as to what provisions were necessary in such an establishment, not only for conducting every variety of physical exercise, but for securing at the same time the comfort, improve-

3

ment, and health of the students.   While the building has
a large main hall for general exercises, it has numerous
other rooms of different sizes, most conveniently constructed, located, and arranged, for all needful purposes.
It has provisions for every kind of bath, with abundance
of water — cold and warm.   From a careful inspection of
the apparatus, equipments, conveniences, etc., it would
seem that every thing was here provided that is possible
for the highest welfare of the students.

Some account should be given of the measurements
of students.   We can not here go into details, but only
make a general statement.   Upon the admission of every
new class to college each student submits to some sixty
different measurements of his body and its parts, such as
weight, height, lung capacity, girth of chest, arm, etc.,
and an exact record of all these measurements is kept.
These examinations are repeated every year, and since
they commenced over twenty - five hundred different
students have been thus measured.   In a report just published by Doctor Hitchcock, of twenty-five years' experience in gymnastics, there are twelve tables, containing the
summing up or results of these measurements.   They
show an immense work — that thousands and thousands
of figures have been employed to obtain these results.

While these measurements have a present value to
every student, and to the cause generally, in the course of
time they become invaluable in aiding to settle some problems in vital statistics connected with physiology, biology,
and anthropology.   The statistics thus gathered will bear
fruit through successive generations.

Remarks on Early Education.

———•———

STUDY, or mental exercise, when properly pursued, is
productive of health. This result harmonizes not
only with the laws of physiology, but it is confirmed by
all experience. It is a well-established fact that educated
men, as a class, reach a greater age than an equal number
of persons whose minds have not been much cultivated.
This increased longevity is obtained not simply because
such individuals possess the means or knowledge of taking better care of themselves, but that a certain kind and
amount of exercise of the brain has a salutary influence
upon all other parts of the body. The question naturally
arises, What are the proper guides in this matter, or what
are the limitations to which mental exercise can be safely
carried? In order to arrive at any definite or satisfactory
knowledge upon the subject, it becomes necessary to understand some of the laws of the physical system with
their relations to mental development, and, on the other
hand, what is the reciprocal influence of mental exercise
on the body.

That there is a most intimate and interdependent relation between the body and the mind, requires here no discussion. If there is any advantage on either side, the
former has it, for there may be found great numbers in
every community possessing good physical development
and comfortable health, without remarkable qualities of
mind; but distinguished abilities and attainments, combined with a well-developed and healthy physical system,

are by no means so common. The evil most to be depre-
cated is that these mental accomplishments are so frequently obtained at the expense of, or rather by breaking
down, the physical system, followed by a life of feebleness
and suffering, and, not unfrequently, by premature death.
This depends much upon the training and habits early
formed in the family and the school-room.

One of the most important conditions for securing good
health and long life is a well-balanced constitution. For
illustration: the human body may be compared to a complicated machine, composed of many nicely-adjusted parts.
Every mechanic will say at once, that the more perfectly
balanced all these are kept while in operation, — so that
each part performs well its own distinct work,— the better
and longer such a machine will run without getting out of
order. Here the "wear and tear" comes on all parts
equally, or just in the manner or extent which its construction or adaptation designed. It is just so with the
human body. The health and lives of thousands are sacrificed every year by some single point — frequently very
small — of defective organization or violation of physical
law, when all other parts of the body are perfectly sound,
and would have performed their respective functions for
many years. Where one member suffers, all suffer, and
where one dies, however premature, all die, no matter how
sound and healthy. Such is the law of Nature.

It is true much depends upon the soundness of the
original constitution, as affected by the laws of hereditary
descent and other causes ; still the organization, by proper
care and training, may be materially modified and changed
in early life. Those parts that are weak and defective
may be strengthened and more fully developed ; while
great pains should be taken that no particular class of
organs, or any one of the temperaments, should become
unduly predominant. It should be borne in mind that

this change or improvement, if ever made, must be done when the various organs of the body are in a state of growth. And here comes in an important consideration, viz., the laws of growth ; according to physiology this is the natural order : First, the cellular tissue, then the muscular, the cartilaginous, the osseous, and last, the nervous. Inasmuch as the muscular is the moving power of all the other tissues, it becomes of the highest importance that its proper exercise and healthy development should be carefully attended to in childhood and youth.

Whilst the various organs of the body are going through this process of growth, a great change of organization, by particles, or cells, as they are termed, is constantly taking place. There is not only the change required in the daily supply and waste for carrying on the natural functions of life, but, in addition, a certain amount of nutrition, rest, exercise, etc., is absolutely necessary for promoting the healthy growth of the system. Hence, in that period of life in which Nature intended the body should receive this growth, it is exceedingly important that each and every part should have its natural growth, in order to secure the right balance of organization. The supplies by way of nutrition, sleep, and exercise, should be abundantly adequate, not only for support, but for growth, so that every organ should have its due share in kind and amount.

Among the four great agents, viz., food, sleep, air, and exercise, employed in promoting the growth of the system, that of exercise, in its various forms, is, in some respects, the most important. While the exercise of some portions of the body is in a great measure involuntary, — such as that of the digestive, arterial, and respiratory organs, — there are other portions whose use depends more upon the volitions of the individual. Such are the muscular and nervous tissues, the exercise of which has a powerful influence for good or evil over all the other organs. Too much can scarcely be said in favor of the proper exercise of the muscles in their influence upon physical strength and health, as well as mental accomplishments.

In discussing mental exercise and its effects, there are several points that should constantly be borne in mind.

First. That, following the natural order or law of growth, the brain and nervous system come last.

Second. That the brain, for healthy growth and exercise, requires relatively a greater amount of nutrition than other parts. It is allowed by physiologists, that in a normal state, about one-third of the blood should go to sustain the brain ; and thus, in this way, one-third of the vitality of the system is consumed.

Third. No kind of exercise uses up the vital energies or exhausts the system like that of brain-work.

Fourth. That the brain, when properly nourished by good blood and refreshed by sleep and recreation, will work to far better advantage than when impoverished, weakened, or overtaxed.

Now, what is the effect of prematurely overworking the brain upon the organ itself, or upon other parts of the system ? It is evident that when all the tissues are in a growing state, they require a larger amount of nutrition. If the brain is excessively or unduly exercised, it demands and receives more than its legitimate share, so that the other organs must suffer, — it may be the stomach, the heart, the lungs, or the muscles. Their natural growth and development are therefore checked, resulting at first in some very slight weakness or derangement, not perceptible at the time ; but gradually it increases, and ultimately becomes a most serious complaint or dangerous disease. It may be the very first approach of indigestion and dyspepsia, or the starting point of some one of those endless weaknesses and diseases connected with debility, anæmia, consumption, etc.

The Education of Girls, Connected with Growth and Health.

———•———

ONE of the chief causes of failure in education is the want of fixed principles as guides. In all matters appertaining to the welfare of the mind or body, we should always have some definite principles to direct and guide us. The Creator has established such laws in the human system for the proper development of every faculty of the mind, as well as organ in the body. It is the province of physiology to unfold the nature and character of those laws, in their various applications.

There is one general law in this science which should be better understood. In the whole process of education a most important change is constantly taking place in the physical system, which is but little noticed, viz., *growth*. There is not only the regular law of supply and waste going on to support life, but, in addition, Nature demands that provisions should be made for the increase or growth of every part of the body. This law commences operations with life, and continues to adult age, though the changes which Nature makes at particular periods are greater than at some others. An observance of this law of growth is of the highest importance in the whole course of education.

But before noticing these laws and changes, let us inquire if Nature has not established some general or normal standard to which we may always appeal. In considering any subject there are great advantages in having before us some perfect model or pattern, by which every part can be tested.   In the organic world, we believe, there is everywhere such a standard, though it may be difficult to find perfect examples of it.

In physiology there is a normal standard, and it consists in *perfection of structure and function;* that is, that every organ should be sound in formation, and capable of performing its legitimate office.   Thus in the human body, all its parts must be sound and well developed, and each must perform its own proper function, without interfering with that of others.   The human body may be compared to a complicated machine, where every part has a specific work to do.   Hence will be seen the importance of having the balance or harmony kept up, so that the "wear and tear" shall come equally upon every part.   The wisdom of such construction and operation is very obvious.

The most thorough researches into both the sciences of anatomy and physiology demonstrate that there is such a standard of organization ; and upon this foundation is based the law of health and life.   It is a *normal* standard, a universal law; and the nearer all parts of the body can approximate toward it, the greater will be found the aggregate amount of health, and the longer human life.   In the growth and changes, therefore, that take place in the body, it is of the highest importance that this standard or law should be kept constantly in view.   Among the Greeks and Romans, where physical organization was made of great account, we find models set forth corresponding to this standard.   In the Apollo Belvidere and the Venus de Medici we find the most complete illustrations of development in all parts of the body.   Experience and observation taught the Greeks and Romans that such standards of organization, of all others, were the most desirable ; but the principles of physiology not only demonstrate the fact, but explain the reasons for it, and the modes by which it is obtained.

The organization here described furnishes the foundation not only for the laws of health and long life, but presents the true standard of beauty, where symmetry, proportion,· figure, and outline are exhibited in their highest perfection. There is still another principle involved in this same organization, more important than either or all of the others, — that is, the law of multiplication and continuance of the race. A volume might be written upon each of these topics, and the object of making these general statements here is more particularly to show what may be the effect in education of deviations from this *normal* standard of education. It is true, we shall find no perfect examples, — only approximations towards them, composed of an almost endless variety and character. If these deviations from the normal standard are very marked, they are attended with more or less unfavorable results. This depends very much upon what organs, or class of organs, are included in the deviation.

PHYSICAL ORGANIZATION.

It may be said we can not change the physical system materially; that is beyond the power of individual choice or agency. To a certain extent this is true. There is a fixed type, or there are marked features, in the organization of some races, like those of the Jews, which continue for ages. It is so, though to a much less extent, in some families, where their leading characteristics are transmitted for generations. But in both these instances the natural relations are generally confined to the same classes, for just in proportion as this relation extends beyond kinship or race, will there be changes in the sameness of organization.

While the principal features grow out of the laws of

inheritance, radical changes must require two or three generations. Still, many important changes do take place in the life of an individual. If the laws of growth and change were better understood and observed, it would be found that we have far greater power over the physical system, in *development*, than is generally supposed.

It is an established fact, that every part of the body is constantly changing, so that in the course of seven years it is estimated that the whole system is entirely changed, not a particle of the same matter remaining at the end of that time. And although these changes are carried on according to certain fixed laws in chemistry and physiology, we have the power, to some extent, of directing and modifying their results. The two principal agents in effecting these results are nutrition and exercise. Careful examination shows that these two agencies are controlled very much by our own choice and power.

GROWTH OF THE BODY.

The human body is made up of infinitely small cells, and the various changes it undergoes are very properly called cellular development. The principle of "waste and supply" is here admirably brought into exercise. While nutrition from food and air is continually furnishing the means, a set of vessels is provided to carry off all waste matter. These vessels or carriers are called the capillary system, and though at times they are exceedingly busy, yet they never cease their work night or day.

The cells composing the primary elements of the body consist of different orders, and vary in form and size. The bones, muscles, ligaments, nerves, brain, etc., are all built up by cells, and are nourished by cells, formed from food and absorption outside through the lungs and the skin. Different kinds of food are, to a certain extent, designed to make a particular class of cells; for instance, some form muscular fiber, others nerve tissue, and others adipose matter. The capillary system, which is the agent in these changes in cell-life, is composed of exceedingly small blood-vessels, and is distributed everywhere through the body. They act as connecting links between supply and waste, as messengers carrying nutritious cells, and removing those that are waste and decaying. In the whole process of digestion they act as agents, after the food has passed through certain changes, in carrying the nutrition to its place of destination, and then removing the waste matter. They form an important connection between the arteries and veins, exchanging pure blood for that which has become impure, extending to the purification of blood through the lungs.

Without entering too minutely into physiological questions, our object is to show briefly what are some of the laws of growth and change in the human system, and that these are, in a great measure, dependent upon human agency. While we may not at once be able to understand all the points or principles involved in the subject, enough may be seen to show how important they are, and that they should be far better understood.

While we can not draw the line between the kinds of food, as to their exact adaptation to build up this or that tissue, it is well understood there is a great difference, and that selections can be made with special reference to developing the muscle, or strengthening the nervous system, or increasing the lymphatic temperament. If all children possessed the same organization throughout, the same kinds of nourishment would be adapted to all alike ; but as there are exceptions to this general rule, greater care should be exercised in such cases. The time will come when this whole subject will be better understood, and the laws of nutrition, as applied to all cases, will be more carefully observed.

Connected with the laws of growth and support, the prompt removal of all waste matter, or the secretions, is highly important.   Unless it is done, this effete or decaying matter poisons the parts surrounding it, or re-enters the circulation, and becomes the cause of much disease. Nature has made ample provision for this work, but its operations are often thwarted by human agency. For illustration, we may refer to the importance of cleanliness of the skin, or to the normal action of the alimentary canal.   Another illustration may be given in attending to the healthy action of the lungs, that they be not only supplied with an abundance of pure air, but that the impurities generated by internal action should not be retained by compression, or want of exercise; and when once expelled, the smallest particles should never be allowed, if possible, to re-enter the lungs again. In the early stages of education, when the individual has no knowledge on the subject, and is entirely dependent upon a parent or teacher for guidance, it is highly important that these rules be applied, for they are then most needed and will do the greatest good.

In providing suitable food for the body many things must be taken into account, and this is far more important to young persons while growing than to those who have reached adult life.   Attention should be given to the demand of Nature in the adaptation of food, that all parts of the body may receive those kinds most appropriate for their growth and development.   Besides, it is not the mere kind or quality alone, but there must not be deficiency in quantity, neither should it be taken in excess. Then there is the preparation, — the cooking part, — which is vastly more important in the case of the young than is generally considered.   The health, growth, and constitution of children depend greatly upon the preparation of their food.   The manner and times of taking food should receive careful attention; the food should be taken slowly and be well masticated; should be consumed at regular, set times, — at intervals of five or six hours, and nearly in equal portions, unless at the last meal, which should be light, — care being taken to preserve always a good appetite.

In the application of the principles here presented there are several important considerations. From the age of five to twenty the growth and development of the body should receive special attention, whereas the practice at the present day is reversed. In the matter of education the mind absorbs all attention, but the claims and interests of the body are regarded as of too little consequence. What are the teachings of physiology on this subject? The principles of this science, and the lessons taught by experience, should be the guides. It is very obvious that the brain, upon which all mental manifestations are dependent, embracing so small a portion of the physical system, should not receive all the attention. From the age of five to twenty Nature provides especially for the growth of the body, so that all parts of it should obtain at twenty, or soon after, a healthy and complete development. After this period, there is no natural growth of the body as a whole, but changes may occur in different organs, and especially the brain. During all these years the main object of Nature in the organization seems to be preparatory work, — growth, training, development, strength, etc.

From this general law, we should infer that no one part of the body should be exercised at the expense of other parts, so as to produce a premature development. It is clear that if the exercise is carried beyond the laws that regulate a healthy growth, and interferes with the normal development of other parts of the body, the result must be exceedingly injurious. Physiology teaches *unmistak-*

ably that the normal standard is based upon a sound, well-balanced organization, and the nearer to it is the approximation in the development of all the organs of the body, the larger the amount of health, the longer the life, and the greater the human achievement and happiness.

THE TEMPERAMENTS.

This principle may be illustrated by the doctrine of temperaments. For the sake of convenience, we take the most simple division, viz. : 1, The *Muscular*, or motive, made up of the ligaments, etc., and muscles generally; 2, The *Sanguine*, including the heart, lungs, arteries, veins, etc. ; 3, The *Lymphatic*, composed of the lymphatics, absorbents, and glandular system ; and 4, The *Nervous*, including the brain and nerves throughout the body.

Now, the more evenly balanced these several temperaments are, the more healthy and perfect is the organization. Each organ is better able to perform its own specific duty, and, of course, there are greater harmony and less friction in their operations. In such an organization there is far less chance for weakness or disease to obtain a foothold.

If there were slight deviations in the balance of the temperaments, it would make but little difference in the health or strength of an individual; but if any one of these temperaments becomes altogether predominant, it will be accompanied with serious disadvantages : especially if this should happen to be the muscular or nervous, for these temperaments constitute the leading agents in the development of the organs embraced in the other two. The muscles involve the motive-power, — the law of exercise, — which lies at the foundation of growth and health. The nervous temperament includes the brain, the organ of will and thought, which, of course, must have a powerful influence in directing and shaping the development of the whole system.

It may be said we have no power to change these temperaments, that we can not change or mould the organs of the human body at will. It is true there are bounds or limits in the changes of organic matter, beyond which we can not go; but then, by commencing early in life, and persevering in the use of proper means, there is abundant evidence that great changes can be effected. The size and strength of certain parts or organs in the body have been, in many instances, materially changed. Illustrations could easily be given, where the size of muscles has been greatly enlarged, and where the power of the lungs and other organs has been surprisingly increased. The fact is, scarce any attempts have been systematically and thoroughly made for the improvement of the young in this direction. It will never be known what can be done in this way until the trial is actually made; and before any radical changes or reforms can here be effected, we must understand better the evils now existing. We can notice only the more obvious of these evils, with a few suggestions as to their remedies.

EVILS AND REMEDIES.

One of the most encouraging signs of the times is that the attention of the public is being directed more and more to physical improvement. There are undoubtedly serious objections to some of the ways in which this interest is manifested, especially as connected with athletic sports and games. The matter here may be carried too far for the physical and moral interests of those engaged in them. Where this improvement is most needed, is in early training in the family, combined with an educational system. Physical improvement should become a leading object, both in the family and in the school; and, through all the stages of education, the culture of the body should go hand in hand with that of the mind. It should be made to apply especially to those who need it most, whose organization is weak or defective, — where some parts are imperfectly developed or not well balanced, and there is lack of strength and harmony of function. There should be in all schools a system of gymnastic or physical exercises of some kind, wisely adapted to the varied wants of the pupils.

In advocating a more strict observance of the laws of health and life, and objecting to the present modes of education, it should be distinctly understood that no one department of mental culture, no particular mode of teaching, neither the higher education of women, nor co-education, are here singled out for criticism; neither is it intended to oppose or object at all to female education; but, on the other hand, we advocate the highest possible mental culture for girls that is compatible with their whole organization, that harmonizes with both the physical and mental systems. This constitutes the only sure basis or foundation for all true culture, and its laws are the certain tests of its correctness and success; for, guided by these laws, there is no theory, no experiment, no failure.

In making application of the principles here laid down, reference will be made more especially to girls, as both in the family and in the school they are less provided with the means for physical development than the boys; while, considering the nature and objects of their organization, it is far more important for girls. Within a few years the education of girls has been pressed with great energy, especially in New England. In cities and large villages, girls are sent annually to school from five years of age to sixteen or seventeen, with the exception of ten or twelve weeks' vacation each year. In small towns and rural districts the amount of schooling is less, perhaps from half to two-thirds as much as in cities. While great stress is laid upon the kind and number of studies, and the standard is raised in the mean time higher every year, scarcely any attention is given to the growth and development of the body. With rare exceptions, there is no system of gymnastics or calisthenics provided in schools for girls, and, generally speaking, no regular and systematic exercise that is adapted to promote their highest physical welfare.

In examining the effects of such a course of study the laws of physiology must be our guide. If we should consider, in all its bearings, the relation of the mind to the body through life, it would seem as though the latter should receive as much attention during these ten years as the former. It is a question whether, by such a course, the great object of existence might not, in a larger measure, be secured. It is a fact that many young people who grow up in the country, with very limited schooling, excel in scholarship and attainment those trained in the schools of the city. It is also a fact that, where the half-time system of schools has been conducted a long series of years, the pupils (working half of the time) have made as much progress in learning as those attending school all the time.

That we may obtain more definite views of the effects of education, as now conducted, let us consider some of the physiological changes produced by it. The muscles and the brain constitute the two leading forces in the human system, and may be represented by the motive and nervous temperaments. It is of the highest importance that these two temperaments should be fully developed and made prominent in the growth of the body; otherwise, the organs included in the other two temperaments will never attain their proper growth and complete development.

The muscles constitute by far the largest portion of the body; they grow only by exercise, and become strong and healthy only by much exercise. Thus they receive their proper share of nutrition, increase in size and strength,

4

and gradually obtain that most important quality, — fitness for work and power of endurance. This exercise of the muscles must commence early, and be continued year after year, so that the fibers of the muscles, by repeated extension and contraction, become hardened and toughened ; their possessor can then work, and hold on without being tired, will have what is called *great power of endurance.*

On the other hand, where there is deficient exercise and a want of proper growth and development of this temperament, the muscles are pale and weak, soft and flabby ; they have not sufficient vitality and strength to carry on, in a healthy and vigorous manner, the machinery of the whole system. The muscular temperament, when well developed, receives a large supply of blood, and constitutes the leading agency in causing a free and equal circulation of blood through the whole system ; whereas, when the muscular power is weak, there is a great tendency to frequent congestion, especially in the internal organs, which prepares the way for much weakness and many diseases.

Besides, this muscular power in large supply is needed to obtain good blood by a more vigorous action of the lungs and stomach ; no one thing is more important for good health than a free and equal circulation of the blood. This muscular power can be obtained only by a great deal of exercise when young, and no substitute by friction, stimulants, or other human devices can be found to replace it. Individuals deficient in this power labor through life under great disadvantages.

Again, we have stated that, when in the course of education, and as a result of it, there is a great predominence of the nervous temperament, and a lack of the muscular, the internal organs of the body do not stand so good a chance for growth and devolopment. As a consequence, these same organs suffer in weakness and greater liabilities to disease ; the lungs, from consumptive complaints ; the stomach, from indigestion and dyspepsia; the bowels, from costive habits ; and the reproductive organs, from a variety of weaknesses and diseases. The heart also suffers in its action, for the want of muscular power, and in case of weaknesses and diseases in different parts of the body, it can not force the vital currents so well throughout the whole system.

The weaknesses and diseases of all these organs originate more or less from the want of muscular power, and then this defect comes from neglect of the kind and amount of physical exercise which should have been taken while the body was in a state of growth and development. But an excessive cultivation of the brain or the mind has, directly and indirectly, done its full share in producing these evils.

The fact here stated brings us to one of the most serious evils in the present modes of education. While it cultivates the mind and stores it with knowledge, training the mental faculties to their highest extent, and capacitating them for the greatest happiness, it developes at the same time an organization, which, unless it has health, the means and ability to be gratified, becomes susceptible of immense suffering, both of body and mind. It may be said that such a result can not be prevented, especially in some cases, but alas! they are altogether too common, and are likely to increase more and more unless some radical reforms are effected.

In confirmation of the statements here made we summon the two following witnesses. Miss Elizabeth Blackwell, M. D., who pursued, some years since, a thorough professional course of study in Philadelphia, and is now a successful practitioner of medicine in London, says, "We need muscles that are strong and prompt to do our will, that can run and walk in-doors and out of doors, and convey us from place to place as duty or pleasure calls us, not only without fatigue, but with the feeling of cheerful energy. We need muscles so well developed that shall make the human body really a divine image, — a perfect form, — rendering all dress graceful, and not requiring to be patched and filled up and weighed down with clumsy contrivances for hiding its deformities; bodies that can move in dignity, in grace, in airy lightness or conscious strength; bodies erect and firm, energetic and active; bodies that are truly sovereign in their presence, expressions of a sovereign nature. Such are the bodies we need; and exercise, the means by which the muscular system may be developed, assumes then its true position as of primary importance during the period of youth. It is the grand necessity to which every thing else should submit."

Mary J. Studley, M. D., connected a long time with the State Normal School for Girls at Framingham, Mass., says: "It has been my privilege, for more than twenty-five years, to be intimately associated with young women, either as a teacher in the school-room in the earlier years, or as medical practitioner or teacher of hygiene during the latter ones, and every day's added experience only confirms me in the position I have occupied from the first relative to the various forms of nervousness which characterize our sex. That position affirms that the best possible balance for a weak, nervous system is a *well-developed muscular system*. Weak, shaky, hysterical nerves always accompany soft, flabby muscles, and it is a mournful fact that the *majority of the young women* whom I meet in schools are notably deficient in muscular development."

True Basis of Education.

ACCORDING to physiology, all education should be pursued in harmony with the laws which govern the brain and the physical system. Formerly very little attention was paid in education to the condition of the body or the development of the brain; and even at the present day far less than should be to those great physical laws which underlie all mental culture. The lives of a multitude of children and youth are sacrificed every year by violating the laws of physiology and hygiene, through mistaken or wrong methods of mental training; besides, the constitution and health of a multitude of others are thus impaired or broken down for life. Nowhere else in society is a radical reform needed more than in our educational systems. Inasmuch as the laws of the body lie at the foundation of all proper culture, they should receive the first consideration. But in educating the boy or girl, from the age of five to fifteen, how little attention is given to the growth and physical changes which necessarily occur at this important period of life! The age of the child should be considered, the place of schooling, the hours of confinement and recreation, the number and kinds of studies, together with the modes of teaching, should harmonize with physical laws — especially those of the brain.

The system or mode of treating *all children as though their organizations were precisely alike*, is based upon a false and unnatural theory. Great injury, in a variety of

ways, results from this wrong treatment ; in fact, injuries
are thus inflicted upon the sensitive organizations and
susceptible minds of young children, from which they
never recover. That many of our most independent and
clear-headed educators themselves express so much dissat-
isfaction with the working results of our schools affords
evidence that something is wrong in the present system.
As we contemplate the great improvements made in education for the last thirty or forty years, and are surprised
that educators were content to tolerate the state of things
then existing, so will the next generation, when still
greater and more radical changes shall have been introduced, look back with astonishment at this generation, and
wonder that it was so well satisfied with its own methods.
When our educators become thoroughly convinced that
physical development as a part of education is an absolute
necessity, — that a strict observance of the laws of physiology and hygiene is indispensable to the highest mental
culture, then we shall have vital and radical changes in
our educational system. The brain will not be cultivated
so much at the expense of the body, neither will the
nervous temperament be so unduly developed in proportion to other parts of the system, often bringing on a train
of neuralgic diseases, and exposing the individual to the
most intense suffering which all the advantages of mental
culture fail, not unfrequently, to conpensate.

The more this whole subject is investigated, the more
reason we shall find for making allowances or some distinction in scholastic discipline with reference to the differences in organization of children, and for adapting the
hours of confinement and recreation, the ventilation
and temperature of school-rooms, the number and kinds
of studies, the modes of teaching, etc., to the laws of the
physical system. But another and still more important
change must take place. Some time — may that time be

not far distant — there will be a correct and established system of *mental science*, based upon physiological laws ; and until this era arrives, the modes and methods of education must remain incomplete and unsatisfactory. The principles of this science, in the very nature of things, must rest upon a correct knowledge of the laws and functions of the brain ; and until these are correctly understood and reduced to a general system, all education must be more or less *partial, imperfect,* and *empirical.* While the old theories of metaphysicians are very generally discarded, they still have practically a powerful influence in directing and shaping our educational system and institutions. In the selection and arrangement of studies very little attention is paid to the peculiar nature or operations of the various faculties of the mind, or the distinct laws that govern their development and uses. For illustration — instead of educing, drawing out, and training all the mental faculties in their natural order and in harmony, each in proportion to its nature or importance, the memory is almost the only faculty appealed to in every stage of education ; and this is *so crammed and so stuffed* that frequently but little of the knowledge obtained can be used advantageously. Instead of developing the observing faculties by "object teaching," appealing to the senses of sight and hearing, — those two great avenues of knowledge, — or giving much instruction *orally*, we require the scholar to spend most of his time in studying and poring over *books*, mere *books*. The mind is treated as a kind of general receptacle, into which knowledge almost indiscriminately must be poured, — yes, forced, — without making that knowledge one's own, or creating that self-reliance which is indispensable to its proper use. In this way the brain does not work so naturally or healthily as it ought, and a vast amount of time, labor, and expense is wasted — nay, worse than wasted. From this forced and unnatural process there often results not only a want of harmony and complete development of all parts of the brain, but an excessive development of the intellect with the nervous temperament, and not unfrequently an irritability and morbidness which are hard to bear and difficult to overcome. And not unfrequently it ends in a permanent disease of the brain, or confinement in a lunatic asylum.

When we take a careful survey of the various discussions and diverse theories on this subject, considered metaphysically, and then compare them with the great improvements and discoveries in the physical sciences for the last fifty years bearing upon the same subject, the change of progress looks mainly in one direction, viz., that all true mental science must ultimately be based upon physiology. Here is a great work to be performed, and when accomplished it will constitute one of the greatest, most valuable, and important achievements that was ever wrought in the history of science. A large amount of positive knowledge has already been accumulated on this subject by various writers, but a great work, by way of analysis, observation, and induction, and of further discoveries as to the functions of the brain, remains to be completed.

Normal Standard of Physiology.

IT is admitted that Professor Huxley is the highest living authority on matters pertaining to physiology. The following table, prepared by Professor Huxley, defines the constituent elements that compose a perfect human body. It describes exactly not only all of its principal parts, but what supplies it must have, from day to day, to preserve it in a healthy state. This table reads as follows:—

"A full-grown man should weigh 154 pounds, made up thus: muscles and their appurtenances, 68 pounds; skeleton, 24 pounds; skin, 10 1-2 pounds; fat, 28 pounds; brain, 3 pounds; thoracic viscera, 3 1-2 pounds; abdominal viscera, 11 pounds; blood which would drain the body, 7 pounds. This man ought to consume per diem: lean beefsteak, 5000 grains; bread, 6000 grains; milk, 7000 grains; potatoes, 3000 grains; butter, 600 grains; and water, 22,900 grains. His heart should beat 75 times a minute, and he should breathe 15 times a minute. In twenty-four hours he vitiates 1750 cubic feet of pure air, to the extent of 1 per cent.; a man, therefore, of the weight mentioned, ought to have 800 feet of well-ventilated space. He would throw off by the skin 18 ounces of water, 400 grains of solid matter, and 400 grains of carbonic acid, every twenty-four hours; and his total loss during the twenty-four hours would be 6 pounds of water and a little above 2 pounds of other matter."

This description represents a harmony or balance of human organization which, we believe, has practically very

important bearings.   We have, in this description, set
forth to a certain extent both the anatomy and physiology of
the body — the structure in the fore part, and the function
in the latter part.   This organization may very properly
be considered the *normal standard* of the human system,
— that it is represented here in its best estate.   While
we may not, perhaps, find perfect examples like the organization here described, we find all manner of approximations towards it.   Still the standard remains the same,
and upon it are based, we believe, certain great physiological laws, which are fundamental and vastly important.
Some of these laws we propose to notice briefly in this
article.

I.   *The Law of Health.*— In analyzing this table we
might almost scientifically figure out the exact changes
which cause disease.   There must be, in the very nature
of things, one kind or type of organization more conducive
to health than another.   Admitting this fact, there must
be an organization of the body far better adapted to secure
perfect health than all others.   What, then, must be its
type or character?   What must be its anatomy and its
construction?   Is not that the standard which consists in
a perfect development of all the organs of the human
body, so that there shall be a perfect harmony in the performance of their respective functions?   By referring to
the table, it will be seen at once that a change in the
weight or measures pertaining to any part of the body
will make a radical change in the type or standard set
before us.   If you change any one of these factors, you
destroy the harmony or balance in the whole organism.
If the structure is changed, it impairs just so much of its
functions.   This constitutes the entering wedge of disease.
The particular kind or character of the disease must
depend upon what organ or part of the body is changed.
By referring to the table we find certain directions given

as to the support of the body. If there is a failure to carry out these directions, or if there is any material change in the character of the supplies, disease may not at once be produced, but the vital forces of the system may be lowered, or some weakness started. The first changes may be slight in their character, but lead to serious results. Some of the gravest diseases originate from the most trivial causes.

There are, it may be said, different *degrees* of health. This fact is very obvious. What makes the difference? It is not because this or that organ alone is sounder in one person than another. It is not simply because one person takes so much better care of himself than another, though this makes quite a difference. If we bring together all the causes or reasons, we shall find that the secret consists in the fact that the constitution of one is more perfectly and evenly developed; that there is greater harmony and completeness in the performance of the functions of all parts of the body. There must, therefore, be a *general law* regulating this whole matter of health — some standard, some type of organization, better than all others. As far as figures can explain it, we find it described in the table at the head of this article. In other words, it consists in that type or standard where every organ in the human body is perfect in structure, and where each performs perfectly its own legitimate functions. In some respects the body may be compared to a complicated machine, so thoroughly and perfectly made that the "wear and tear" will come equally upon every part according to the design in its construction.

Closely connected with, and legitimately following, this condition of things, we find Nature has established another law, viz. : —

II. *The Law of Longevity.* — Is there not some standard or model laid down by physiology itself, that shows why, in certain cases, life should be protracted to a great age? It does not depend upon food, climate, locality, race, or care, though all these may have much to do with it. Is there not an *internal factor* more potent than all these? One of the great secrets, we believe, not only of good health, but of long life, consists in the harmony or balance of organization. This must apply both to structure and function. The leading vital organs should be not only sound, but well balanced. The principal forces in carrying on the functions of life may be summed up under these heads: respiration, digestion, circulation, assimilation, and secretion. Each of these departments must be well sustained in order to secure long life.

But, aside from any theory, or opinion, or argument, what are the actual facts — what do we find in the organization of those persons who have reached a very great age? No tables or statistics can be given from postmortem examinations of such cases, because attention has not been turned in this direction. But, from the physical description of a great number of very aged persons, and from careful observation also of a very large number, we have always found that a most striking harmony or balance of the physical system prevailed. In great longevity there is uniformly found remarkable consistency or evenness in the mental, moral, and social elements of character. These traits originate from a sound and well-developed brain. This organ plays a very important part in securing longevity.

There is another argument in favor of this law of longevity — that the extremes in physical or mental development seldom reach a very great age. It should be borne in mind that the law of longevity here advocated constitutes the golden mean, or balance-wheel, between these extremes. For instance: the defective classes, such as the insane, the idiotic, the deaf-and-dumb, the blind, etc., are not, as a body, long-lived; neither are dwarfs or giants, nor persons approximating such organizations, very long-lived.

There is another very important factor in longevity, — that is, *inheritance.*   Scarcely any fact on this subject is more firmly established than that the ancestry, the family, or stock, has much to do with long life.   Seldom, if ever, do we find a person reaching a great age without some one or more persons in the ancestry have reached a great age. What, then, is the peculiarity, or type of organization, here perpetuated?   What are its elements that make life so long?   Do we not find that they consist in a sound, healthy structure of every part of the body, and that there is a remarkable balance in all the organs, and a harmony of functions?   So universally is this essential element found in persons long-lived, that we question whether a single exception to the rule can be found.   This leads to another application of this normal standard of physiology, — that upon it is based

III.   *The Law of Heredity.* — For centuries there has been more or less interest on this particular topic.   A large mass of facts have been gathered upon the subject, and physiologists now generally admit that there must be truth in this matter of inheritance.   Within a few years the interest has greatly increased.   In the case of domestic animals the principle has been reduced almost to a science.   With some changes or modifications, the same principle which has been so successfully applied to the animal creation will apply to human beings.   But before there can be great advances on the subject, we must understand heredity better, we must have some general law or principle to guide us.   What we need more than any thing else, is a general principle or law, by means of which all the facts or knowledge of this kind can be classified and reduced to some system.   It is impossible to make any great advances or improvement upon this subject of heredity without such a guiding principle or standard of appeal. In the facts or phenomena of Nature there must be some general law or principle to guide us in understanding them and improving upon them. All science makes progress only in this way.

While there may be different factors and secondary causes in producing many of these hereditary phenomena, if the primary cause or starting point could be ferreted out, we might find it to extend back several generations. All the general principles of science, when traced back to their origin, are based upon Nature in its best condition. And the nearer we can go back to a perfect physical organization, the less peculiarity, eccentricity, or defect shall we find. It may be we can not explain or understand all the causes of strange or different phenomena or character ; it does not disprove but there may exist a general law somewhere. It is true, there have been different theories and speculations in accounting for hereditary influences, but we do not believe that they can all be explained so satisfactorily upon any other law or hypothesis as upon the one here stated — that is, upon a perfect development of anatomy and physiology, or in other words, that all the organs in the human body shall be so constructed that there must be legitimately a healthy performance of all their functions.

There is another important test in favor of this normal type of physiology — as far as the human body is concerned, it presents the true *standard of beauty.* Man was created with a sense of taste and love for the beautiful, which, cultivated and perfected, might find objects in Nature capable of gratifying this taste to its fullest extent. Now there must be a type or model for man, which in form, proportion, size, fullness, outline, is more beautiful than all others. Is not this the same standard that

Grecian and Roman artists have attempted to imitate in statuary? Has it not, in all ages and with all nations, attracted attention? Why should it not constitute the basis or foundation for most valuable laws? But the most important law of all, involved in this physiological description, remains to be stated, that is

IV. *The Law of Human Increase.* — This law virtually controls all the others. With a change here, the conditions of health, of longevity, and of heredity would necessarily be more or less affected. It is, in fact, the starting-point, the ground-work of most important inquiries that can be raised connected with physiology. All that we can here do is to state briefly what this law is, what some of the evidences in support of it are, and what are some of its applications.

In the first place, there is no universal law of population, that is generally admitted as such and referred to as authority. Nearly one hundred years ago Malthus established what he supposed a general principle to regulate population, and his theory prevailed for fifty years or more. It is discarded now by nearly all physiologists, as well as most writers on political economy. It is rare to find now any prominent writer advocating the doctrines of Malthus. The theories of Herbert Spencer on this subject have probably, at the present day, more influence than those of any other writer. The views of Spencer, unlike those of Malthus, are based upon physical organization, but are not so strictly physiological as the law here proposed. The foundation, the ground-work, of the law we advocate, is based solely upon anatomy and physiology in their best estate. There are other factors, such as food, climate, exercise, and other external agents, but these are secondary.

That this law may be distinctly understood, we will describe, as briefly as possible, what is meant by it. It is based upon a normal or perfect physical standard of the human system, where every organ of the body is complete in structure and performs fully all its natural functions. This principle implies that the body is symmetrical, well-developed in all its parts, so that each organ acts in harmony with all others. According to this principle the nearer the organism approaches that standard and the laws of propagation are observed, the greater will be the number of children, and the better will be their organization for securing the great objects of life.

On the other hand, if the organization is carried to an extreme development in either direction, viz. : a predominance of the nerve tissue, or of a low animal nature, the tendency in such families or races is gradually to decrease and ultimately to become extinct. Thus people enjoying the very highest civilization, or living in the lowest savage state, do not multiply rapidly. It is well known that the families in Europe belonging to the nobility or aristocracy, whose nerve tissue has become predominant by intermarriage from generation to generation, do not increase much, and not unfrequently these families become extinct.

A similar result has followed the intermarriage of relatives, from the fact that the same weaknesses or predispositions to disease are intensified by this alliance. On the other hand, in case these relatives have healthy, well-balanced organizations, — it may be they are cousins, — they will abound in healthy offspring, and the stock may improve, and not deteriorate, from the mere fact of relationship. It explains a principle that has long been employed in the improvement of domestic stock, under the terms, "breeding in and in " and "cross-breeding."

Physical Development.

———•———

AS this phrase, "physical development," has been selected for a running topic, or central point, in this
work, it seems proper that some explanation should be
given of its meaning. The word "physical" is well understood as referring to material things, but the term "development" may be interpreted in different ways. It is
defined by Webster as the "unfolding or unravelling of a
plan or method, or series of progressive changes"; implying that there had been certain preparatory work before
the thing itself had reached its present form. When the
term is used connected with physiology, it is understood,
in one sense, to signify the size or form of any part or
organ of the body. Then, it has a relative meaning,—
that is, when one part or organ is compared with another.
There is still another sense in which the word may sometimes be used,—that is, where we have a standard or
model set before us, by which a comparison is made. This
last use of the term development is very important; but
much here depends upon what is this standard or model.

Many years since, after much study and reflection, we
became convinced that there must exist in physiology a
type of organization which very properly might be designated a normal standard. Upon examination, we find a
great variety or difference in organization, as far as health
or disease is concerned. The term "normal," when applied to physiology, is understood to indicate a healthy

5

state, or freedom from disease. As there may be different degrees of health and one type of organization better than any other, we would call the best a normal standard. It presents, therefore, a model or rule to guide us in all our investigations connected with this science. In all such studies there is great advantage in having a law or rule constantly in mind, and, in case of difficulty or doubt, to which we can appeal. Without some law or standard, it is hard to make advances in any science or department of knowledge. In physiology this becomes doubly necessary when we want to improve organization and remove causes that perpetuate disease. The more this subject is investigated, the more important it will be found that such a course is necessary.

In a preceding article will be found a standard where the constituent elements were furnished by Professor Huxley. Here the different parts composing the human body are obtained mostly by weight. The pattern described might be considered that of an average man, or a physio_logical type of the highest order, and might very properly be reckoned as one of normal standard.

Two very important experiments are being carried on in this country to obtain correct standards or models of the human body. One of these is conducted by Dr. D. A. Sargent, of Harvard university, and the other by Dr. Edward Hitchcock, of Amherst college. This is done by taking an immense number of measurements of college students, including every part of the body. Doctor Hitchcock has been taking these measurements for over twenty-five years, and has examined nearly three thousand different students. He is still prosecuting his inquiries, and collecting an immense amount of figures and tables. The greater the number of students examined, and the larger the quantity of statistics on the subject, the more perfect will be the standard. Doctor Hitchcock is here working out a problem which, in time, may be turned to a most valuable account.

In Professor Hartwell's report of physical training in American colleges and universities, we find this sketch of Doctor Sargent's views: "The object of physical training with us is not to make men active and strong as much as it is to make them healthy and enduring. Perfect health implies a condition in which all parts of the body are properly nourished and harmoniously developed, — in which the vital organs are sound, well balanced, and capable of performing their functions to the fullest extent. The researches of the physiologists have shown that, whenever a certain organ or class of organs becomes relatively too large or too small, causing a want of balance or harmony in their action, there is in every case a far greater liability to disease. It is in imperfect, ill-balanced organizations that we find the greatest amount of sickness and the greatest number of incurable diseases. It is the weak spots caused by inheritance, acquired by exposure, close confinement, overwork, etc., that invite disease and death, even though the rest of the system be in perfect condition. To attain a perfect structure, harmony in development, and a well-balanced organism is our principal aim."

Doctor Sargent has been engaged in this work many years, and has made more examinations than any other person. Now having a physiological standard of health before him, in examining a person, he can more readily find weaknesses and defects. Thus having ascertained these points in individual cases, the question is, What can be done to remedy these evils? This opens a most fruitful field for study and experiment. It is found that by special training, and the use of apparatus adapted to each case, these weaknesses and defects can be very much modified, and in some instances overcome. By this means the health and constitution of such persons become also much improved.

Doctor Sargent has for years been doing here a good work, and finds his labors in demand in many different institutions. Among these are several female colleges and seminaries, where such training may be turned to a most useful account. "To attain perfect structure, harmony in development, and a well-balanced organism," is a movement in the right direction ; aiming to secure what may be truly designated a "Normal Standard of Physiology."

This increased interest in physical culture in our educational institutions is an encouraging sign. As it bears its own fruit, which must be evident to all, this reform must go forward. No greater or more important reform can be carried on in behalf of education than improving the physical development of students.

In addressing the alumni of Harvard university, not long since, President Eliot made this remark : "Now, everything depends with us, and in the learned professions, *upon vigor of body*. The more I see of the future of young men that go out from these walls, the more it is brought home to me that professional success, and success in all the learned callings, depends largely upon the vigor of body, and that the men who win great professional distinction have that as the basis of their activity." If careful inquiry on this point were made in respect to the graduates of other colleges, we believe the truth of this remark would be abundantly confirmed.

The Law of Longevity.

THE subject of longevity has always attracted much attention. The art and means of prolonging life were frequently made the themes of discussion, long before the real structure and functions of the most important organs in the human body were discovered. But as the principles of physiology have of late years become better understood, new interest has sprung up in relation to all matters pertaining to health; and the inquiry is very generally raised at the present time, what are the best means of preserving life, and thus securing that great boon, longevity? Now, may there not be a general principle or law, grounded in physiology, which may serve as a guide in these matters, and help to illustrate and explain all minor facts or secondary considerations? Is there not some standard or model established by Nature herself, to which we may always appeal, and by which all doubtful questions here may be tested? From our knowledge of the laws of Nature, as well as of the principles of science generally, we should naturally infer that there must be found in physiology some such general law, or such standard. Several years since, after somewhat extended observation and no small amount of reflection and reading, we became convinced that there existed a general law of population, or increase, as a fundamental principle in physiology, and that this same law of propagation (subject to certain conditions) extended throughout the whole animal and vegetable kingdoms. If such a law in Nature does exist, it might be inferred that

it would have some connection with the greatest amount
of health and longevity.

Law of Propagation. — This law may be briefly defined
thus : it is based upon a perfect standard of organization,
or consists in the perfectionism of structure ; or, in other
words, that every organ in the human body should be perfect in structure, and that each should perform its legitimate functions in harmony with others. Taking this, then,
as a standard, we have a great law or principle pervading
all organic matter, that furnishes a guide by which all deviations from this model, and the manifold changes that
follow, may be explained and understood. While this law
is subject to certain conditions, as food, climate, exercise,
etc., all these act as secondary agents or factors. They
may modify the operation of the law, but can not change its
nature or general character.

Evidences in proof of such a law may be deduced from
physiology itself, from pathology, from the laws of hereditary descent, from the effects of intermarriage of relations,
from facts gathered in the history of different families, and
changes in numbers, as applied to distinct classes, races,
and nations. But without dwelling upon these points we
maintain that the organization upon which this law of propagation is based, presents also the only true standard in
physiology for the greatest amount of longevity — of health
— of physical strength and happiness, as well as of beauty
in form and outline.

Law of Longevity. — But it is proposed to consider here
only the application of this law to longevity. By this term
is meant long life — the greatest duration of human life,
whether in isolated cases or in large numbers. Where,
then, are these cases found — what is their character —
and what are the facts attending them ?

In the first place, it is very evident that long life is not
dependent alone upon food, nor upon climate, nor upon

exercise; neither is it found in any one locality, nor with any one people, nor in any particular station; neither where great riches or excessive poverty prevail. It is sometimes found in the city, but more generally in the country.

All must admit that some of these conditions are very important, and that good health and long life must depend greatly upon the manner in which the relations between the various parts of the system and these external agents are carried on. But after all, may there not exist a general law in the body itself upon which these depend? If we had perfect standards of organization around us upon which this law is based, its truth would be more easily demonstrated; but instead of such, we have only approximations, and these in almost endless variety and form. In order that we may have a clearer and more definite understanding of the foundation of this law, let us carefully examine its physiological conditions. Every animal organization is complex — is composed of many distinct organs. Each organ has a specific work to do, and in its normal state must do so much and no more. Now in the healthiest and most perfectly organized structure, all these separate organs are found not only in a perfectly healthy condition — each one performing its own normal functions — but well balanced and working harmoniously together. In this state "the wear and tear," or the demands which Nature makes to support life and carry on its operations, come upon all these organs alike, each according to its own nature, without infringing upon that of any other.

The Human Body Compared to a Machine. — In the promotion of health and longevity, too much stress can not be attached to the importance of preserving this harmony or balance of organization. In some respects the human body may be compared to a perfect machine, made up of many complicated parts. How different the working or running of such a machine from that of one imperfectly constructed and unequally balanced in all its parts! The one seldom needs repairs, the other frequently. The one will last as it were for an age ; the other becomes almost useless in a short time.

It is so in reference to the human system. Whenever a certain organ or class of organs becomes relatively too large or too small, causing a want of balance or harmony in their action, there must be in the very nature of the case far greater liability to disease. Accordingly, it is in persons possessing this imperfect, ill-balanced organization, that we find not only the greatest amount of sickness, but that which is most obstinate and fatal. How often it happens that some slight derangement or trifling weakness operates as the entering wedge to the most serious diseases! It is the weak spot caused by inheritance, or developed by exposure, where disease finds its germ or starting point, though all other parts of the system are in a perfectly sound condition ; and not unfrequently life is terminated by a single organ, or even some part of it, giving out, when all the other organs might have performed their healthy functions for many years.

We dwell upon the importance of this harmony or balance of action in the vital forces, for it is the great secret of good health and long life. It is a cardinal point in the law of longevity, as will appear from a more full sketch of its foundation.

Perfect Structure and Harmony of Function. — It is upon this perfect structure or anatomy of the body, combined with the normal action of all its physiological functions, upon which we base this law of longevity. It is true we have no such perfect standards or models of human organization now existing, but only approximations towards them. Still the law may apply to such as we have, just as well as the general law of gravitation or attraction to the smallest-sized bodies. We can readily conceive of such standards, and how the same law that governs them may be applicable to their representatives of whatever grade or character.

All the pains, the weaknesses, and the diseases of the human body are but the result of deviations from this normal state ; and all the means and agencies employed for the preservation of health and life look towards restoring this standard. It is well known that there are influences constantly operating to produce changes both in the structure and functions of the system. Some of these agencies have their origin internally; some act entirely external to the body, and others operate by what are called laws of heredity. By some of these influences the physical system is improved and perfected, but by others the deviations from a healthy standard are increased more and more. Probably the most powerful of these forces is that of inheritance. This agency constitutes a very important element in the law of longevity. All writers upon this subject place this condition as first and foremost — that one of the almost indispensable requisitions for long life is good healthy stock, or long-lived ancestry. For it has been found by universal observation and experience, that the representatives of such stock live the longest, and that very seldom, if ever, are found persons of great age originating from feeble and short-lived ancestry.

Law of Inheritance. — Now what is the secret of this transmitted power that conduces so much to longevity ? May there not be some general principle or law involved in these changes from hereditary influences, which may aid us in explaining the why and wherefore ? We know well the effects of such power, but what is the explanation — what is the philosophy involved ? Under the law in Nature that "like begets like," and that when the producing forces are sound and healthy it is found that their offspring will partake of the same character, and that under favor-

able circumstances this may be continued for several generations. Sometimes there is an improvement in the
stock ; but not unfrequently a deterioration, especially after
three or four generations.

Now what is the peculiarity or type of organization here
perpetuated ?   What are its elements or constituents ?
What makes it long-lived ?   Do we not find that it consists in a sound, healthy structure in every part of the body,
and that there is a remarkable balance in all the organs
and harmony of functions ?   We venture the assertion that
such will be found the character of this organization in
every instance, and that there are no exceptions to the rule.
Does not this, then, afford evidence that there is a general
law in Nature conducive to longevity, and that this law is
based upon that organization which is most perfect, and all
of whose functions act most harmoniously ?   Let us apply
the rule to such individuals and families reaching a great
age, that have come under our own observation.   For many
years we have verified the fact in numerous cases, and have
never found an exception.

There is another point of view whereby this law may be
tested.   Certain physiological conditions have been laid
down by some writers as sure indications of longevity.
These conditions embrace the healthy performance of the
functions of all the leading organs of the body, and may be
summed up under these heads : Respiration, Digestion,
Circulation, Assimilation, and Secretion.   Where all the
vital forces connected with each of these departments of
physiology are found to operate regularly and vigorously,
they are thought to be the sure indications and precursors
of longevity.   Now what does this imply but soundness of
structure and harmony of function ?   Let any one of these
fail in the least of performing its part, and all suffer.   Does
not this view of longevity, then, furnish strong evidences
in favor of the law which has been set forth in this paper?

Signs of Longevity. — There is another class of facts which have an important bearing upon this question. These are what are denominated the physical signs of longevity. There must be a symmetrical development of the whole body. The head must not be too large or too small. The neck must not be too long or too slender. The chest must be well developed, but the abdomen must not be too large. The whole body must be well proportioned, not too tall nor too short. No class of organs must be too predominant ; or, in other words, the temperaments must be properly mixed or blended; especially the nervous and the sanguine, possessing more of the vital organs, must not be very conspicuous. There are some minor signs, such as the voice, the teeth, the color of the eyes and the skin, the quality of organization, etc. ; but when we sum up all the foregoing signs, do they not clearly point to a harmony or balance of all the organs of the body, and thus confirm the truth of the law of longevity as here advocated ?

There is a large body of facts also connected with the cure and prevention of disease, that has a direct bearing upon this subject. All sound medical treatment, and means for the promotion of health, operate in harmony with this great law of longevity. They aim to restore the normal structure and healthy functions of every part of the body.

In all works treating of longevity great stress is laid upon the influence of climate, food, air, water, exercise, etc. Statistics show that, while the extremes of either heat or cold are not conducive to long life, a moderate climate, in countries where the changes of temperature are neither too great nor too sudden, is decidedly favorable. But even here there must be a strict observance of hygienic laws. In relation to the right kinds of food and drink, pure air, healthy localities, dwellings, employments, etc., however important, they are all secondary agencies, and operate under and in harmony with one general law.

Mental Hygiene. — But there is still another class of
facts differing from any of those mentioned, that has a
powerful influence upon longevity, viz., the influence of
mind upon the body. Mental training, a well-balanced
mind, a cheerful, contented disposition, and temperate
habits are, with rare exceptions, found indispensable. Now
these presuppose an harmonious development of the whole
body, and particularly of all parts of the brain. For it is
impossible, we believe, to obtain the qualities here mentioned in a high degree without these two conditions.
And the nearer this development approaches that standard of organization upon which is based the great law
of longevity, the greater will be not only the aggregate
amount of health, but the longer the duration of human
life. This statement will be found abundantly verified
in the history and character of persons who have reached
a great age.

This interdependence of body and mind is becoming
every year better and better understood. It is found that
the relations of the mind to the body, and of the various
states and changes of physical organization to the mind,
have a powerful influence upon health. And the more
marked and abnormal the differences in this relation, the
more striking are the effects. If, then, health is so dependent upon the state and relation of these two agents,
the duration of human life must be most sensibly affected
by it. And we venture the assertion, that the more thoroughly this particular feature of the subject is investigated, the more important and far-reaching will be found
the influence of these reciprocal relations. The evidences
derived from this source will go far, we believe, towards
proving that Nature has established a certain harmony or
equilibrium of action between the body and the mind, and
the more perfect that development and harmonious the
performance of their respective functions, the nearer is the

approach to that standard of organization upon which is based the law of longevity.

This view explains, in part, why the average age of man has been increased by education, and that the greatest longevity is found among nations most highly civilized. In confirmation of this remark, a distinguished writer says : " That type of civilization in which the efficiency of the community and of the individual is greatest, in which there is the most harmonious action between the body and the mind, the greatest happiness of the greatest number, the least excessive expenditure with the least luxury, where regularity and temperateness are innate characteristics, will be that state of civilization most favorable to longevity." It is scarcely necessary to say that such a type of civilization could not exist without well-developed physical organizations generally, and an harmonious action of all the mental faculties.

Another well-known writer on this subject, after enumerating among the prerequisites to longevity, temperate and regular habits, a cheerful and contented disposition, says there must be not only an equilibrium of the mental faculties, but a descent from long-lived ancestors, a tranquil and happy temperament, a general symmetry of physical conformation, and harmonious proportion of all the different parts and organs of the body.

Numerous quotations might be cited from other authors, and many additional facts might be gathered from various sources in support of this theory of longevity ; but our present limits will not permit. Perhaps the theory of one writer should not be passed by unnoticed, inasmuch as it may be thought to have some resemblance to the one here presented.

Theory of M. Flourens. — M. Flourens, in a very elaborate treatise, maintained that man ought, by virtue of his natural constitution, to live a hundred years, and that this

natural term of life is abridged only by his own improvidence, follies, and excesses. The length of human life he
attempts to establish by the law of growth and by analogy,
viz., that every animal will live, on an average, five times
the period of his growth. Thus, as it is found by anatomy
that it takes, on an average, twenty years for man to reach
his perfect growth, especially the bony structure, the limit
of life would be one hundred years. Flourens held that
neither climate, nor food, nor race, nor any external condition, had much to do with the duration of life, but this
depended almost wholly upon the natural constitution, and
the intrinsic vigor of all the organs of the body. But he
does not define very clearly how this natural constitution
is based upon the anatomy and physiology of the system,
nor attempt to show what are its laws and relations to the
external world. We all know that climate, food, and other
external agents have a powerful influence upon the development and preservation of the body. One great defect
in his theory is, that he does not point out distinctly the
great laws of health and life as based on physiology and
external nature, which extend not only through individual
existence, but are universal throughout creation. As to
the question what is the natural period of human life, provided all the conditions are favorable, perhaps he is not so
much out of the way, though the testimony of most writers
would place the limit somewhat less. Flourens presents
us no standard of organization as a perfect model of imitation, and upon which the great laws of health and life
must be based. If we take into consideration the structure
and functions of the human body, — the design of its existence and its adaptation to external objects, — there must
be certain relations and fixed laws that govern in all these
matters. For illustration : there is a fixed law that exists
in the relation of pure air to the healthy functions of the
lungs. It is so in reference to all other parts of the body.

Now it is in the summing up of all these laws, as applied to a perfect organization, that we find the law of longevity. All the great laws of Nature, that are fixed and universal, are invariably found based upon her works in a normal state, or in their most perfect development. As in painting and statuary the artist has constantly in his mind an ideal model, — a typical standard which no living beings have ever reached, but only made approximations to, — so in physiology it is easy to conceive of a standard which represents an organization in its highest state of development. It was with reference to the making up and arranging the constituent elements which enter into such a standard that led the most profound physiologist in our country (Professor Draper) to make this remarkable statement: "The approach to precision in these hypothetical constants will in all times be a measure of the exactness of physiology, and, it may be added, also of the practice of medicine. The time is at hand when such a typical stand. ard must be the starting-point for pathology, and no rational practice can exist without it. *The passage of physiology from a speculative to a positive science is the signal for a revolution in the practice of medicine.*"

Advantages of the Law of Longevity. — The question may very properly be asked, Supposing there is such a law of longevity, what are its advantages? We answer, many and great. It is not a mere speculative theory, or vague hypothesis, that can not be comprehended or applied to any practical purpose. It harmonizes not only with all the well-known truths of physiology and pathology, but is sustained by all the agencies employed by Nature or Art for the protection and preservation of life. In fact, it is that great general law established by the Creator himself for perfecting and prolonging the life of every human being, of which all minor laws are a part and parcel. It holds up before us that perfect form and image in which man was created, and presents an embodiment of those laws and conditions with which we must comply in order to secure the greatest amount of happiness and the longest duration of life.

With such a standard constantly before us, shall we not make greater efforts to conform to it, than if we had no such conception? Besides, by means of understanding the various deviations from this perfect standard, we obtain a better knowledge of the infirmities, the liabilities, and the weaknesses of the human system. It presents a new stand-point from which to survey the causes of disease, as well as the agencies employed for its cure and prevention. It gives us a clearer and better understanding of the principles of hygiene and sanitary law, and enjoins the absolute necessity of observing them, if good health and long life are to be secured. It shows that all the changes which occur in the human system are subject to law; that disease, of whatever type and character, or wherever found, is a violation of law; and all treatment and remedies, whether provided by Nature or Art, must be viewed as agents or means to repair the injury.

But there is one use to which this law may be applied of incalculable value; we refer to life insurance. This is becoming an immense business; scarcely surpassed in interest and magnitude by that of any other in the country. From the best sources of information it is estimated that there are over five hundred thousand, or half a million, of lives insured in over two hundred different companies, and the amounts invested and at risk would startle one not accustomed to figures. The largest proportion of this business has sprung up within twenty or thirty years, and what is singular, the larger the business and the wider its expansion, the greater the changes in its management, and the more uncertain are its results. We should naturally suppose that time and experience would give permanence and stability; but what a sad spectacle is presented by the rise and fall of so many life insurance companies — some of them, too, after many years, apparently, of successful experience! What a history of wrecks, losses, and disappointments does it exhibit! Scarcely can a parallel be found in the history of any other incorporated business in the country.

In the examination of any organic structure, with reference to forming an estimate of its continuance, we must understand correctly its nature and construction, as well as the laws that govern its action. If it is made up of many parts or distinct organs, we must comprehend fully their relations to each other and to external objects. But in order to make the best use of such knowledge, and form an intelligent estimate of results, we want some general law or standard of appeal, which shall be applicable to the whole. To any one acquainted with the earlier history of the different sciences, it is well known what great advantage was found when a large body of facts or amount of knowledge had been obtained; that by the discovery of a general principle, all these facts and this knowledge could be more systematically arranged and satisfactorily explained. It is somewhat so in applying this law of longevity to life insurance, though it may be subject to many conditions and can not be reduced to mathematical accuracy.

Prerequisites of Longevity. — Without explaining again this law and its conditions, let us briefly notice some of its applications in determining the prospect or continuance of life. All the essential elements or prerequisites for longevity may be conveniently arranged or summed up under three distinct heads, viz. : constitution, inheritance, and obedience to law.

First. It furnishes the examiner for life insurance with a standard of organization, with which the constitution of

G

all persons examined may be compared, and which will assist in forming a correct judgment of their soundness, or in detecting the physical deviations from a normal standard; then, what are the liabilities to disease, and what the prospects or probabilities of life. Without such a standard or guide we have no general rule to test the soundness or strength of the constitution. It must depend very much on opinion merely, which, of course, will vary according to the differences of judgment in different individuals. With such a model constantly before us as Nature has furnished, we can understand more exactly and fully the relations which all parts or organs of the body sustain one to another, as well as to external nature; and then we can calculate or forecast far better the changes to which they may be subjected. The more of such knowledge we possess, the more accurately can we estimate the continuance or prospect of life.

Second. *Long-lived Ancestry.* — All writers upon life insurance lay great stress upon inheritance, or a long-lived ancestry. This has been found by universal experience to be one of the prerequisites — in fact, an indispensable condition of long life. Now, why — why is this so important? What are the reasons? What does it mean? What is the rationale of it, or what lessons does it teach? Does it not clearly and distinctly imply, that if there is any truth in this power of inherited organization for long life (the more perfect the organization, the greater the power), there must certainly be found, somewhere in nature, a great general law of longevity? The influences of hereditary descent have as yet received but little attention, compared with their importance, even by the medical profession ; and before they can ever be thoroughly understood it will be found, if we mistake not, that there exists in physiology, as a fundamental principle, a general law of propagation, and as a part and parcel of the same will also be found this law of longevity. In the matter of life insurance, a thorough knowledge of these hereditary influences is of the utmost importance.

Third. *Conditions of Health.* — Obedience to law. This has a very wide application, including all the physical laws and relations of body and mind. The better these laws and relations are understood, and the more strictly all are observed, the greater will be the amount of health, and the longer human life. But in order to effect this most successfully, the conditions of good health must first be fully understood, such as pure air and water, wholesome diet and drink, healthy vocation and residence, regular exercise and sleep, temperate habits, right mental and moral culture, with a cheerful, contented disposition.

Balance Among the Physical Functions. — Under this heading a friend has just placed in our hands the following testimony from Dr. H. C. Wood, professor in the University of Pennsylvania : —

" To make it possible to live to a good old age, the several vital organs must be approximately equal in strength. The man of ordinary physique, who possesses this fortunate balance of power, will, in all probability, outlive an athlete, whose development has been unequal. Excessive strength in one part is, in fact, a source of danger. All overdeveloped muscular system invites dissolution, because it is a constant strain upon the less powerful organs, and finally wears them out. Death, in the majority of cases, is the result of local weakness. It often happens that a vital organ has been endowed with an original longevity less than that of the rest of the organism, and its failure to act brings death to other portions of the system, which in themselves possessed the capabilities of long life. The fact of having succeeded in life, with the satisfaction and comfort it brings, contributes to the prolongation of existence, while failure, with its resultant regrets, tends to shorten it. In old age the organs possess less elasticity to meet and overcome such strains as can be invited with impunity in youth. Hence the old should be spared strains."

Duties of Medical Men.

A DISTINGUISHED French savant, not long since, in
speaking of the personal influence of educated men,
made this remark: "We must not live here in this world
without leaving traces which shall make us to be remembered by posterity." This statement, on the first impression, might seem rather presuming, if not arrogant; but
upon more reflection, we believe it indicates a purpose
and determination that are commendable. And if this
remark can properly or justly be applied to any class of
educated men, it is surely to the members of the medical
profession.

Let us inquire, then, what are the reasons and what are
the circumstances which justify such a statement. What
are the relations, particularly, which medical men sustain
to the public and posterity? No man in this profession
should rest satisfied with present attainments, nor allow
himself to settle down in a mere routine course of study
and practice. Not only in justice to himself, but for the
public welfare, as well as for the good of posterity, he
should do something more; otherwise, he leaves no permanent traces or marks by which he may hereafter be
remembered. In the present state of medicine there are
certain subjects that require of its cultivators special
attention. At no former period in the history of medicine
have the causes of disease been so carefully scrutinized,
or in which greater advances have been made in this
direction. Never was there a time when the natural laws

of disease could be studied to so good advantage, — the primal cause and real nature of every distinct disease, — and to learn that the whole mystery concerning disease arises mainly from our ignorance. Never was there a time in medical history when we might understand so well the exact place and adaptation of medicine in the treatment of disease; to avoid the extremes, on the one hand, of over-medication, and on the other, the rejection of all medicine.

The profession has been inclined, we think, in the past, to use altogether too much medicine, both in quantity and variety, which has produced serious evils. This practice has misled the members of the profession in placing too much confidence in the virtue of medicine, and not enough in the recuperative powers of Nature. It has also inspired the community with such unbounded confidence in medicine alone, that it is difficult to change their views and instil into their minds the value of sanitary influences. On the other hand, the physician who carefully studies into the causes of disease, and the laws that govern it, is apt to lose faith in medicine, and is liable to take radical ground against the use of all medicine. A well-balanced, progressive mind will avoid either extreme.

Here is a grand opportunity for a physician to do good service, to make his influence felt and known both in the profession and in the community. It is in the direction of improvement where advanced views in respect to medical practice and the good of society can be entertained. In this way the medical student may establish landmarks which will cause him to be long remembered. If all diseases are the result of violating the laws of health and life, the symptoms are the indications or outward signs of this violation, and medicines are employed to aid in removing or allaying the disturbances.

Drugs can never mend broken limbs; neither can they

restore the system to a perfectly sound and healthy state, when its inherent laws have been repeatedly violated. Nor can disease be rationally and successfully treated, until its causes and nature are better understood. Medical practice can never be reduced to a true system of art or science until the causes and laws of diseased action are better understood.

A most important field of study is opened in this direction, in which an inquiring mind may do immense good. It is true, researches have here been made by many students, but there is still room for further explorations. Men in the profession, both in England and in this country, are waking up more and more to inquiries of this kind. Marks will here be made which will be remembered.

Closely connected with this improved rational treatment of disease is a most interesting field opening for study and influence, — that is, the prevention of disease. Formerly *cure* was the supreme, if not the leading, object of the physician, but the time is fast approaching when *prevention* will be the watchword.

What can you do to prevent disease ?- This requires study, thought, and a sincere love of humanity. There is here a great work for the medical profession, and it is the duty of its members — a duty which they owe to themselves and to the public — to engage earnestly in such work. This can be done in a variety of ways, — in daily private practice and by seizing every opportunity to enlighten the community on the subject by speech or by the pen.

Here the thoughtful and progressive physician can exert great influence and establish landmarks which will long be remembered. In fact, no one question is coming before the public of such vital importance and magnitude as this prevention of disease under the head of *sanitary science.* The principles of this science are based on the laws of physiology. While there has been no change in the functions of physiology, there have been discovered new or far more important relations, as it respects health, between this science and objects external to the body.

There is a natural adaptation between certain organs in the human system and outward objects, which, if brought into their true relation, are healthful and normal. Pure air, pure water, and pure soil, representing the externals on the one hand, and on the other, physical exercise, proper diet, and suitable clothing, representing the body, —these constitute the ground-work, the pillars, of sanitary science. These apply to human life in all its connections and aspects. They include nearly every agency and influence operating upon the growth and development of the body, or upon the occupation and pursuit of every individual, or upon the food, drink, and home life, his place of residence, and dwelling. They cover the whole ground of his education, —physical, intellectual, and moral, —and his religious being, whether they harmonize with the laws of physical organization.

But the principles of sanitary science have a far wider scope than the individual, — they extend to the family, to the school, to the state, and to the church. They take cognizance of all the agencies connected with those great centers of power and influence, in order to determine their sanitary effect on the human system —whether healthful and normal, or hurtful and injurious.

In each one of those centers the principles of sanitary science as based on physiology must be brought to bear with far greater force and effect than at any former period. There are questions involving the highest interests of all those institutions which are comparatively new in their application, and which no man can discuss so well as a thoroughly educated physician.

Among those questions the following may be mentioned: Notwithstanding the family is an old institution, and its lines or objects might seem to be well defined, still we think there are some things here to learn in the application of physiological laws. This institution is based upon physiology as well as revelation. It can be demonstrated, we believe, from the principles of this science, that the human body can not reach its highest development, neither can the race be perpetuated in its best estate, without just such an institution as the family. It can, moreover, be demonstrated from the same source that the primal objects of the family are three-fold: 1, production of children; 2, chastity; and 3, mutual help and company; and, if there is a failure in either one of these, by design or defective organization, it goes so far to break down the family relation.

Closely connected with this institution is another subject of vital importance — that is, the laws of inheritance. No one but a medical man, or thorough physiologist, can understand so well these laws. From proper investigation these will be found to compose a part and parcel of a great law of propagation, which is a general and fundamental truth.

Physiology is yet in its infancy as respects some of its most important applications. The time will come when the duty and agency of man in respect to observing hereditary laws will be far better understood than at the present time. Closely connected with the family is still another question, which has, within a few years, attracted much attention — that is, woman's position and rights. The settlement of some points in this controversy involves certain physiological laws which persons thoroughly versed in the principles of this science can alone discuss understandingly.

Radical changes in woman's position, employment, and

relations in society, must change more or less her organization, and consequently certain mental qualities, which
may affect sensibly the marriage rate and the stability of
the family.

It may be premature to forecast what those changes
will be, or inquire what will be their effects, but there are
physiological problems involved here, we believe, which are
fundamental, and should come under the review particularly of the medical profession. Some of these questions
in dispute can never be settled by the opinions of individuals, nor by the resolutions of conventions, but by great
principles based upon physical organization. Here is an
opportunity for the members of our profession to settle
questions which will long be remembered.

New and important questions also are raised in respect
to schools, which partake of a physiological and sanitary
character. It is maintained that the mind is being educated at the expense of the body, or, in other words, that
the physical development is too much neglected. Such is
the whole course of education in our schools and higher
institutions of learning, that it causes an undue development of the brain and nervous system, while the exercise
of the muscles and other portions of the body is neglected.
Hence there is not that physical strength and energy, that
vital force and power of endurance to meet the demands
which circumstances and society make upon educated persons.

What is most needed in all our educational institutions
is a well-balanced, symmetrical development of the whole
body, which affords not only the greatest amount of physical capital, but the best ground-work for mental labor and
success. To this end the principles of physiology must be
better understood and more generally applied.

Here is most important work for the physician, where
he can make his influence felt and remembered. In our

common schools and educational institutions, there is great need of a more thorough application of the principles of sanitary science. The matter of ventilation, of temperature, and drainage connected with school-houses, and hours for study, physical exercises, and means of recreation on the part of pupils — these involve sanitary laws and demand the special attention of medical men.

Many in our high schools, colleges, seminaries, and universities are constantly suffering from the violation of these laws. There is a large expenditure of means, time, and labor from ignorance on this subject, which is worse than wasted.

The great objects of education are thus defeated; in many cases weakness, disease, and premature death are produced. Here is a wide range for the application of the principles of physiology and sanitary science. It opens a large, a most promising field for medical men to enter and cultivate. Influences and improvements may here be started which will transmit their benefits to future generations.

There is another great center of power and influence — the State — which stands in pressing need of medical knowledge.

In consequence of discoveries in science and the advances in civilization, new questions are coming up for legislation.

It is found that the health and lives of the people must be subjects of legislation as well as property and material interests. To this end pure air, pure water, and a clean soil must be provided. The question of preventing disease on a large scale is becoming an important subject for legislation. Perhaps nowhere can a medical man do so much good as in the halls of legislation, in devising and enacting laws for promoting the health and the highest welfare of the people.

The more thoroughly the causes of disease are understood, together with the laws of health, the more clearly
do we see how far the preservation of health and life depend upon human agency. Hence the need of increased
intelligence on this subject, and also that new and important legislation should take place.

In France, Germany, and Great Britain, the medical
profession have had far more to do than in this country
in government matters, and in the establishment and
management of public institutions.

Members of the profession have held in those nations
some of the highest official positions, have been leaders in
public affairs, and exercised a commanding influence in
every department of society. It may safely be said, we
think, that the medical profession has held its way there
in position and general influence with either of the other
professions, and the chances at the present time would indicate that it was destined to rise still higher and higher.

While the government and the state of society are very
different in this country, and there may be other extenu-
ating circumstances, the medical profession, if we are not
mistaken, ranks below the other professions in public estimation and general influence; it may, in a few
localities, have a leading influence and command much
respect, but we must admit that on the whole the profession in the United States has not possessed that wide
extent of knowledge, those high scientific attainments and
general weight of character which are found abroad.

The question very naturally arises, why this difference
in the standing and influence of the medical profession ?
Some reasons may readily be given. In the nations mentioned, the profession is much older ; its members are better compensated for services rendered ; many medical men
inherit wealth, which gives them leisure ; then in communities abroad there exists, generally, a higher appreciation

of medical skill and knowledge, as well of scientific and
literary attainments.   But, after all, the fault in this
country rests much with the profession itself.   Its members have not been trained with reference to securing
great distinction in knowledge, in science, and education ;
neither while engaged in professional duties have they
been encouraged to put themselves in the way of prominent positions connected with public institutions, or with
the state or national government.   It would seem as
though there had been an admission on the part of the
profession, if not an understanding in the community, that
medical men could not leave their duties for public life,
either to become scholars, authors, or statesmen.

If the promotion of public health or sanitary interests
were made as prominent in our country as they are in
Great Britain, medical men, in far greater numbers, would
be called into public life.   The time is coming, however,
when this will be demanded.   But the responsibility rests
much upon the profession.   Its members must make themselves more thoroughly acquainted with physiology and
sanitary science in their relations to public health and the
welfare of mankind generally.   Its members must not be
so much absorbed in seeking a mere support, or in the
accumulation of wealth, but must listen more to the calls
of humanity, to the demands of the age, and to the good
of coming generations.

It is by such means that the profession will command
greater respect in the community, will exert a more beneficial influence on the public, and leave marks which will
long be remembered.   The leading object of the Academy
of Medicine is expressed in these words: "To extend
the bounds of medical science, to elevate the profession, to
relieve human suffering, and prevent disease."   The suggestions made in this paper are calculated, we think, to
promote every one of the objects here mentioned.   If the

medical profession is elevated to a higher plane and wider influence, the bounds of medical science must be extended in every direction, which, at the same time, can not fail to relieve suffering and prevent disease on a large scale. If the fields suggested in this paper for medical researches and labors could be thoroughly cultivated it would redound greatly, not only to the credit of the profession, but work out most beneficent results. Never were the calls more urgent for such work, never were the times so propitious, never were the prospects of a rich harvest so favorable. Will the members of our profession respond to those calls, honor themselves in this work, and do justice to medical science and the cause of humanity?

Sanitary Science.

Its History, its Relation to Medicine and the Medical Profession.

----·----

*W*HAT is sanitary science? This phrase is comparatively new, but is full of meaning. The word "sanitary," in its derivation and uses, signifies health or healthy, but when combined with science is far more expressive. It means the application of laws or principles for the preservation of health in whatever way they may be employed. As to the use of the term "science" here, the claim can not be justly called in question. To such an extent have these principles been discovered and applied, and so uniformly and certainly have the same results followed, that they may be said to constitute a science — *the science of health.*

It is not necessary that these laws should be understood by every body, and admitted as true, that they may be considered a science ; but if they have been extensively applied by a large number of good judges, and the same results never fail, they constitute, when combined, literally and truly a *science* — as much so as physiology or biology. As both these are comparatively modern sciences, so is that of sanitation, certainly in name and application.

It is only about forty years since this subject began to attract general attention. It started with the establishment of the registration of births, deaths, and marriages, in Great Britain, by Dr. William Farr. While investigating upon a large scale the causes of death, the inquiry naturally arose, What can be done to prevent, as well as to cure, disease? This inquiry, so simple and natural, has resulted in a most surprising advance in the knowledge of the laws of health and life. So rapid and extensive have been these changes that one living during this period can hardly credit them; and never were there improvements taking place faster than at the present day. But the advantages already secured, though great and invaluable, are mere harbingers of richer and more permanent blessings in store. In the progress of this science, every year has signalized the past, that it had a deeper and broader scope, not so much in improving the old methods of work, but in entering into new fields and enlisting new agencies. Its aim is not merely to remove the existing causes of diseases, but to destroy the germs or seeds of disease. It does not stop with preventing this or that contagious disease, or reduce to the minimum the zymotic class of diseases; but when the principles of this science are applied to the fullest extent, they will present the human body so sound and healthy in all its parts as in a great measure to forestall disease.

There is, we believe, a normal standard of physiology, where all the organs are so sound and well-balanced, and where all perform respectively their functions so thoroughly as to afford small chances for disease. This organization represents the highest standard of health, and the nearer the human body in all its parts approximates this standard, the better or higher degree of health shall every such person possess. With this view of physiology, it will be seen that all disease is a violation of law, whether it arose from internal or external cause. As there must be some change in the structure or functions of certain or-

gans in the body for the introduction of disease, is it not
clearly the province of sanitary science to take cognizance
of such changes? If the violations of law can be arrested
or modified in the very first stages, may it not serve to
prevent a vast amount of disease?

There is a sphere higher and broader, where the principles of this science should be brought to bear — that is, in
perfecting the human body. It is well known that there
is naturally a most surprising difference between one individual, or one family, and another, as to good health and
the liabilities to disease. Why should not sanitary science
recognize this difference more, and point out the way
whereby great improvements can be made in the physical
system, and then eradicate, upon a larger scale, the first,
the primary causes of disease? By commencing early, and
with the use of proper means, the organization of every
individual can be greatly improved and made more
healthy; and by a proper application of the laws of inheritance for three. or four generations, human organization
may become so perfected as to diminish a large proportion
of the sickness and disease that exist at the present day.
This is not mere theory nor speculation, but a doctrine
based upon the laws of physiology — laws which should be
better understood. Inasmuch as such a change would be
productive of sanitation in the highest degree, is it not the
province of sanitary science to enter and cultivate this
field? Would it not improve health and prolong life upon
the largest scale and to the greatest number? What
other science or agency can do this work so well? That
human organization can be improved by the laws of exercise, nutrition, and inheritance, there can be no question.
If the highest state of health depends on a normal standard of physiology, in which all parts of the body are perfect
in structure, combined with a harmonious development of
every organ, it is certainly the province of sanitary science

to use all its appliances to obtain that standard. It is no.
more nor less than the same form or image in which man
was created; and the same Almighty power has established laws by the use of which man, in the process of
time, can attain to that of his original creation. The more
thoroughly physiology is studied with reference to sanitation, the stronger is the evidence that man is the artificer
of his own physical well-being. The laws of inheritance
must become the agents of sanitary science; and healthy
offspring must become an object of primary importance.
When the principles of physiology and sanitary science
are both brought to bear in renovating human organization, we shall find that a wise provision is made for the
redemption of the body as well as the soul. We can not
expect this change will be brought about by divine inter-
ference, nor is it left for accident or chance, but the means
and responsibility are wisely placed in the hands and
power of human agency.

In case the body is thus reconstructed—made sound
and healthy in every part—the germs or seeds of disease
will not be found in the system. Here is work for sanitary science on the largest possible scale. In making
these changes, in order to secure the highest standard of
health and to the greatest number, it will be seen that
sanitary science has a great work to do. The whole system of education, especially in early life, must be based
more and more upon the systematic training and development of the body. There are a multitude of evils in the
present state of society that conflict with the laws of
health and life, which sanitary science would remove or
regulate. Then, in all matters pertaining to mental improvement, to the progress of society, to every phase in
civilization and the various developments of Christianity,
the sanitation of the body and of the mind must be paramount to every thing else. In fact, the province of sani-

7

tary science covers the entire life ; not only of every individual, but of the whole human race.   No other subject or
science is of such transcendent importance.   It is in its
infancy, and no comparison can be made between what it
now is and the magnificent proportions it is destined to
attain.

Taking this view of physiology, and that health is its
normal condition, it will be seen that all deviations from
this state, or violations of the laws that govern it, furnish
the causes or entrance of weaknesses, imperfections, and
diseases which afflict the human system.   These changes
may occur from internal, predisposing causes, or from
agents operating externally to the body.   Just at this
point, in these changes of organization from a normal to
an abnormal state, we are taught most important lessons.
On one side we have sanitation and sanitary science ; on
the other disease and its superstructure, medicine.   Just
here start the most powerful and destructive evils that
ever befell the human family.   These evils may be trifling
in their origin, but increase — sometimes slowly, sometimes rapidly — and become terrible in their results.   They
include the whole catalogue of diseases ; their name is
legion.   We dwell on this point, for it is very important
to have clear and definite ideas of disease, its nature, and
cause.   It is simply the penalty of violated law.   There is
no mystery in it ; no visitation of Divine Providence ; no
curse inflicted by some evil spirit.   It is no less important
for sanitarians than for physicians to have a clear and
definite knowledge of disease as well as its cause.

HISTORY OF SANITARY SCIENCE.

Formerly the great object of the medical profession was
the *cure* of disease.   The programme of studies and lectures
in the medical schools was confined almost exclusively to
this one idea.   The term "hygiene" was scarcely to be

found in books, or referred to in lectures. Physiology was comparatively a new science, and some of its most important applications have not been discovered till within a few years. In fact, this science can not be fully understood in all its bearings without combining with it the principles of hygiene.

The study of physiology was formerly superficial, rather than profound; as the laws of health and life are based on this science, these, of course, were not very well understood. Hence there was great difficulty in ascertaining the real causes of disease and the natural laws that governed it. *Health* and *its normal conditions* must be first understood, and disease — its causes and treatment — come afterward. Very little thought or attention was given to the object paramount to all others: health and its requirements. The whole burden of medical studies and lectures was pursued with special reference to disease and its treatment. Thus in the preparation for the practice of medicine, the treatment of disease has so completely absorbed attention that normal physiology and the recuperating powers of Nature have, in a measure, been overlooked. " *Vis medicatrix* " was a favorite phrase of some writers, but very little use has been made of its practical application. Two great evils have grown out of this defective mode of education: 1st, a lack of clear and definite ideas of diseases and their causes; and 2d, a tendency, in the treatment of disease, to resort mainly to artificial means. But within thirty or forty years there has been decided improvement in respect to both these evils.

From 1840–50 several leading physicians in Great Britain, from careful observation and reflection, began to make some changes in their practice: 1st, to dispense less medicine; 2d, to study more carefully into the natural laws of disease; and 3d, to summon to their aid the powerful resources of Nature. Among these physicians were

John Forbes, John Connolly, Andrew Combe, and others. The *British and Foreign Medical Review* was their organ of publication, which attracted much attention. Several works explaining the views of these men were published at that time, and had a large circulation.

From 1840 to 1850 the registrar-general's office for collecting and publishing births, marriages, and deaths in Great Britain became fairly established. This agency has been more influential than any other for creating an interest in sanitary matters. An examination into the causes of death in different localities, and a comparison of the mortality in one place with another, started many inquiries on public health. The annual reports, also from this office, prepared by Dr. William Farr, added greatly to the interest on this subject. About the same period Dr. Andrew Combe, of Edinburgh, published several works on the application of physiology to education and health. These works had a very large circulation, and exerted great influence in directing public attention to the laws of health and life. The writings of Dr. A. Combe were peculiarly calculated to show the advantages of a practical knowledge of physiology for developing healthy bodies, and thereby preventing disease. While the writings of Doctor Combe were based strictly on scientific principles, they were remarkably well adapted, both in style and matter, to instruct the masses.

One of the most distinguished physicians at this time in Great Britain, advocating reform in medical practice, was Dr. John Forbes. In his celebrated paper called "Young Physic," which was published in the *British and Foreign Medical Review*, he made this significant statement: "Redoubled attention should be directed to hygiene, public and private, with a view of preventing diseases on a large scale, and individually in our sphere of practice. Here the surest and most glorious triumphs of medical practice are to be achieved." If this prophecy has not already been fulfilled, it is very evident that, in progress of time, it will be still more abundantly.

As a result of the interest on this subject, a royal commission was appointed in 1857, to inquire into the sanitary condition of the army in England. This commission recommended that not only some regulations should be adopted for protecting the health of the army, but that a school be established for educating army-surgeons, in which "hygiene and sanitary science" should be taught. This was the nucleus or starting-point of that celebrated work on practical hygiene by Dr. Edmund A. Parks. This "Manual of Practical Hygiene," constituting a treasury of knowledge on sanitation, has had a large circulation, and passed through several editions.

The interest in sanitary matters has been steadily increasing in Great Britain among all classes. Its fruits are becoming every year more and more manifest by improved health generally, and by a reduction of mortality, especially in cities. Numerous acts of parliament have been passed in favor of sanitary science. The medical profession and journals generally commend it; and never were its prospects brighter in Great Britain than at the present time.

Perhaps the science has not created so general interest, nor taken so strong a hold, in the United States as it has in Great Britain; but still its history is one of marked interest. Let us notice a few of its salient points. From 1830 to 1840 Dr. John Bell conducted the *Journal of Health*, in Philadelphia, which very ably advocated the principles of hygiene. In 1835 Dr. Jacob Bigelow, in the annual address before the Massachusetts Medical society, pronounced a certain class of diseases "self-limited" in their character, and urged that they should be treated accordingly. This was a marked step in the way of medical

reform, which, with other influences, led to what was
called the "expectant treatment of disease."

In 1842 was issued the first registration-report of births,
marriages, and deaths in Massachusetts, and these have been
continued annually, till we have now the forty-fourth report.
Sanitary science has been greatly advanced by facts and
arguments derived from these reports. Several other
states have followed the course of Massachusetts, in establishing registration-departments. No one agency can do so
much to advance the cause of vital statistics as such regis-
tration-reports. The application and progress of sanitary
science depend much upon a knowledge of vital statistics ;
and the more thoroughly these are understood, the better
for sanitation.

In 1844 Dr. Elisha Bartlett published in Philadelphia a
work on the " Philosophy of Medical Science," and, in
urging upon the profession a better knowledge of the
cause and nature of disease, said: " The next thing to be
done is to find. out the best methods of modifying and *preventing* disease. This is the great mission which now lies
immediately before us ; *this is to constitute the great work
of the next and succeeding generations."* This statement
was made two years before that of Doctor Forbes, already
quoted. Both these men, living in advance of the times,
were distinguished for original thought and independence
of expression ; they have proved themselves true prophets.

In 1860 one of the most brilliant addresses ever given
in this country was delivered before the Massachusetts
Medical society by Dr. Oliver Wendell Holmes. As this
had a direct tendency to promote sanitary science, the address and its reception deserve special notice. At this
time the importance of a more thorough study of *Nature*
in medical practice had been urged on the profession in
previous addresses and other medical papers published.
In pursuing this line of thought, Doctor Holmes expressed

very positive opinions, accompanied with reasons and illustrations, that too much medicine altogether was given by the profession, and that there were great evils arising from over-medication. For this opinion, Doctor Holmes was not only severely criticised by prominent physicians, but denounced and abused, if harsh language could do it. But reaction soon followed this violent attack. The discussion led many physicians to a new and more careful study of the natural laws of disease and the true effects of drugs. Great good came out of this controversy. Doctor Holmes, instead of being injured, gathered new laurels. Many young physicians, seeing the propriety and force of his strictures, struck out a new course in their practice.

The most effective agents of all, for establishing and applying the principles of sanitary science, are boards of health. The first state board of health in this country was formed in Massachusetts in 1869, since which time boards have been started in nearly all the states of the Union. In 1872 the American Health association was organized in New York. This is the most extensive and powerful agency of the kind in this country, and we think we may safely say in the world. It has published twelve large volumes, which contain a greater and more valuable collection of papers on sanitation than can anywhere else be found. The primary object of the association, as stated in its constitution, is the "advancement of sanitary science." A careful examination of the contents of these volumes affords the strongest possible evidence that the association has done a grand work. Here almost every question connected with the science, in all its diversified applications, is found discussed. Some of the papers show great research and an originality of thought which might be elaborated into a volume. Besides its published works, the association has greatly advanced the interests of sanitary science in all the cities and states where it has held its annual meetings.

RELATIONS OF SANITARY SCIENCE TO THE PROFESSION.

While the success of this science depends mainly upon physicians, there is a wide difference in the interest which they take in it, as well as the sacrifices which they are willing to make for it. Let us inquire who, and how many, of our physicians have been actively engaged in this reformatory work. The number, compared with the whole profession, is small, — in fact, is very small. Those engaged in this work are widely scattered, both in city and country, and are generally active with the pen and tongue, so that they seem more numerous than they really are. There are, it is true, great numbers in the medical profession who are kindly disposed to sanitary reform, and speak highly of it in their practice, but, at the same time, are unwilling to make much sacrifice to advance its interests.

Unlike many other reforms and good works, there is a direct antagonism between the interests of this profession and sanitation. The support of this profession depends mainly on the *cure of disease*, not its prevention. Every step in this reform diminishes more or less professional income. There is no trade or speculation in this reform. When a person has spent years in study, and made large investments to secure a livelihood, how can we expect he will sacrifice these interests? There is probably no class of men, engaged in professional or other kinds of business, to whom appeals of so complex and antagonistic character are made for services. The success depends much upon the education and the moral training of the parties.

On one side stands out the highest welfare of the individual and society, in respect to health, while on the other side the physician is tempted to make his own interests paramount to all others.

Let us for a moment consider his position. In choosing this profession the pecuniary considerations were undoubtedly most powerful; and then, in his early preparations and through his whole course of study, compensation for professional services has been constantly kept in mind. The whole drift of medical study and teaching, by sickness or from books, has express reference to the treatment and cure of disease, — not, as we may say, its prevention. Add to this the most implicit faith that all classes generally have in drugs, together with the crowded state of the profession, it will be seen that the physician is virtually constrained to have an eye constantly on his business. It is true that in medical studies, lectures, and books a great deal is said about the charitable aspects of the profession, and that it is always expected to give a large amount of service to the poor.

It is just to state here that the claims of the sick-poor have been most liberally responded to by physicians, and that no other profession or class of men do so much for the poor as the medical profession. But this work of charity has its equivalents : it secures to the physician a stronger hold in the affection and confidence of the people, and, in different ways, tends to increase his business. But to engage actively in means to prevent disease, not simply in one instance, but in case of great numbers, this is very different, — it cuts off directly the support of the physician. Such action is based upon a love of humanity, — of philanthropy, — a higher range of motive than that of giving services to the sick-poor. It appeals to the very highest class of motives, — not simply to save expense and relieve suffering, or improve health and prolong life, but to elevate mankind and increase, physically, mentally, and morally, the sum of human happiness. Such are the legitimate fruits of sanitary science.

Considering the powerful pecuniary interests of the profession, and the disinterested motives requisite to engage in sanitation work, it is rather surprising that so many members of the profession have from time to time engaged heartily in advancing sanitary science. The main object must have been the promotion of health, the diffusion of useful knowledge, and the enlightenment of mankind generally in respect to the laws of health and life. In some few instances it might have been prompted by pecuniary considerations, — the individual holding some official position, or seeking one. But these are exceptional cases. Our state and municipal authorities have made such small appropriations for public health, that the salaries offered to medical men are not numerous or large enough to be very attractive. In this respect Great Britain is far ahead of us. The promotion of the public health has become there a part of her government machinery. The whole kingdom is divided into some fifteen hundred districts, over each of which a medical officer of health is appointed, with salary graded according to the services rendered. Besides this provision, and showing the interest of the government in sanitary matters, there are over one thousand inspectors of nuisance appointed, in charge of as many districts. This inspection proves of great advantage, not only directly in preventing disease, but by dispersing information among the people, they become helpers in the work.

The medical appointments in Great Britain are made on the ground of special training and qualifications for this kind of work, and the same persons are continued in office for years. Thus there is a wide difference between the interest in sanitary science in Great Britain and in the United States. In the former the science receives a powerful support from the government, and a large amount of means is annually distributed among its advocates. Besides, there is on the part of the people more general intelligence on the subject, — a higher appreciation of the benefits of the science, and a more ready disposition to co-operate in carrying on the reform. Though the science has been making advances in these respects in the United States, there is much room for improvement. Our national government is not doing what it ought for public health ; neither are the state or municipal authorities making the appropriations for it which they should.

Most of the contributions to sanitary science here have been voluntary. This reform has been carried forward by men heartily interested in the work, — very few seeking or expecting any remuneration. The reward for such services does not consist in dollars and cents, nor in the plaudits of the multitude, but in "the consciousness of duty done and noble deeds performed."

A distinguished medical writer lately made this remark : "The most important work that sanitarians are doing at the present day 'is sowing seed, which in time will yield abundant harvest.'" And never in the history of medicine was there such a combination of circumstances so favorable to improvement in the practice of medicine. Never before has there been such earnest inquiry made on the part of the profession to ascertain the true causes of disease. It has been found in the moral world that in order to eradicate great evils, their primary causes must be first removed. So in the prevention of disease, the same course must be taken. This accords with the teachings of sanitary science. Leading members of the medical profession have here been doing noble work.

SANITARY SCIENCE AND MEDICINE.

Some twenty-five years ago Sir Joseph Lister, of Edinburgh, made a great discovery for the prevention of disease by introducing what has been called "antiseptic surgery." It had been found, prior to that time, that wounds and

surgical operations were frequently followed by an inflammation which proved fatal.   Surgeon Lister discovered
that, by an application of antiseptic dressings, patients
were more sure to recover from the most dangerous operations.   It is, moreover, found that antiseptics can be
applied to many diseases, as well as to surgical cases,
which checks their progress and aids essentially in the
recovery.   It is now admitted that a great amount of
disease is thus prevented, and a multitude of lives may be
saved.

Again: in this same line of prevention there has been
made, within a few years, one of the greatest discoveries
ever made in the history of medicine, — that some of the
most dangerous diseases are produced by infinitely small
animalculæ, called bacteria, and other micro-organisms.
This subject is now undergoing most thorough investigation in Germany, France, and Great Britain.   If means
can be devised whereby these bacteria can be destroyed,
or their existence eradicated from the system, it will prevent a vast amount of disease.

Again: there seems to be a prevailing impression in
the medical profession that important changes are about
to take place in the treatment of diseases.   This sentiment is foreshadowed in a variety of ways, and many facts
and illustrations might be cited in proof of the same.
The most noticeable instance is the following: Dr. Austin
Flint, of New York, was invited last year (1886) by the
British Medical association to give an address in 1887
before that body.   Doctor Flint died suddenly in March,
but his address, by singular forethought, was found prepared for this occasion, which has since been published.
The very title of the paper is significant, — "Medicine of
the Future."

No physician in the United States could discuss this
subject with greater propriety and force than Doctor Flint,

and, inasmuch as he was to voice the medical profession in this country, before the highest medical body in Great Britain, it shows the importance he attached to this topic in its selection. At the same time, in presenting these views, he must have been pretty well assured that they would be cordially received by the leading members of that association. After recounting in the forepart of this address the changes that had taken place in his own experience in medical practice, he says: "We are entering upon a revolution in medicine. It is bewildering to project the thoughts into the future in order to foresee the changes which will be brought about in the coming half-century in our knowledge of the correction of diseases, and the results as regards their prevention and treatment."

He expresses the opinion that hygienic agencies will be employed hereafter far more than they have been; that the normal conditions of health and the recuperative powers of Nature will receive greater attention, and less dependence will be placed upon drugs and other artificial means. In referring to bacterial etiology, he says: "Here open to the imagination the future triumphs of preventive medicine in respect to all classes of diseases." When the medical profession, says he, "shall employ all the preventive measures possible and the best remedial medicines, disease will be more successfully treated, and the profession will have reached a high ideal position." Alongside of this testimony we will quote the opinions of three distinguished English physicians who have given special attention for many years to sanitary science.

Says Dr. B. W. Richardson: "The influence which sanitation will exert in the future over the science and art of medicine, promises to be momentous. It promises nothing less than the development of a new era; nor is it at all wide of the mark to say that such new era has fairly commenced. With the progress of sanitary science we must expect to see preventive medicine taking the ascendency. With true nobleness of purpose, true medicine has been the first to strip herself of all mere pretenses to cure, and has stood boldly forward to declare as a higher philosophy the prevention of disease. The doctrine of absolute faith in the principle of prevention indicates the existence of a high order of thought, of broad views on life and health, on diseases and their external origin, on death and its correct place in Nature."

Says Dr. Alfred Carpenter : " The science of disease-prevention is destined to alter the whole field of medical practice; to render obsolete much of our present knowledge as to the history of diseases and the measures which are now required for their treatment. The inquiry must come as to how the increase of disease is to be prevented, rather than, having arisen, how it is to be cured. This will apply to every kind of complaint, and will not be limited to any one class."

Says Sir Henry Acland : " In addition to treatment and cure of disease, whatever be the duty of individuals, medical science and art collectively must aim as a whole—1st, At the preservation of health ; 2d, At the averting of disease from individuals and the public generally ; 3d, At rearing healthy progeny for the family and the state by probing the laws of inheritance; and, 4th, At procuring legislation effectual to these ends. It claims, therefore, a voice in moral education as well as physical training. It holds a duty in relation to the diminution of vice, for the sake not only of self-destroying victims, but more for the sake of the innocents whom they ignorantly slay."

It would seem that in the opinion of Doctor Acland sanitary science covers very important ground. This opinion may be accounted for in part from the fact that he has long been a professor at the Oxford University,—has had large experience in educational matters, and understands the full import of physiological laws. If the preservation of health or the prevention of disease is accomplished by improving the organization, a multitude of other improvements follow, and the more perfect the former the greater will be the latter.

There is one method of preventing disease, referred to by Doctor Acland and other writers, which has never received the attention it deserves—that is by the observance of the laws of inheritance. Within a few years this subject has been considerably discussed in the United States and Great Britain, but few seem to appreciate fully the magnitude of its bearings on sanitation. The diseases considered preventable,—of which there are nine or ten—come under the zymotic class, but there are two other classes, called constitutional and local, each larger than the zymotic. Thus far, sanitary science has expended its principal force upon this class; but supposing its agencies could be brought to bear equally upon the prevention of diseases in these two classes, what a vast amount of good it would accomplish! Let us explain. For many years there has been a class of diseases called "Hereditary," because the predisposing causes were inherited,—because they are transmitted from generation to generation, and thus run in families. Now, if those ancestors were free from any taint, or in other words, had perfectly sound and healthy constitutions, the seeds, the germs, the predisposing tendencies of disease would not be transmitted. Let us carry out a little farther this line of argument.

The same kind of evidence which proves that the germs of, or predisposition to, disease are transmitted in a single instance, applies to all others of a similar character; and the legitimate inference is that there must exist in Nature a great general law. Such a law, we believe, exists and is based upon a normal standard of physiology,—a standard for the government of the human body, wherein all its parts are perfect in structure, and its organs harmonious in their functions. This standard of organization constitutes the highest measure of health; is free from all kinds of weakness, as well as predisposition to disease. But, unfortunately, we do not find such organized standards in the present state of society — only approximations; and the nearer individuals or families approach this standard, the sounder the constitution, the less disease; whereas, the further the deviations diverge from this standard, the greater are the weaknesses and liabilities to disease. Here come in the laws of inheritance, — starting not in a perfect, healthy organization, but in conditions of the body where changes of some kind have taken place in the vital forces of the system. To understand and utilize these laws they must be reduced to some system; the distinct relations between the causes and the effects must be traced out, till we find a great general law serving as a standard of appeal, or a regulator to all the minor ones.

There can be no question but that in the inheritance of *morbid tendencies* we have one of the most fruitful sources of disease. This will become more patent in proportion as the principles of physiology shall become better understood in their connection with hereditary influences. Without attempting to describe the various ways in which the seeds of disease, or the predisposing causes, are transmitted from parent to child, we may say they are *manifold,*— in organization or function; in defective or abnormal structure; in the weak or excessive development of this or that organ; in the general want of balance in the organs, and of harmony of function; in the quality of the blood, and the marked predisposition to certain diseases, like scrofula and consumption.

A class of diseases called "hereditary" has existed since the days of Hippocrates, and has always been considered difficult to treat, and much less to cure. Very little attention has been paid to these complaints by sanitarians, as it was supposed they could not be easily prevented. But this is a mistake; they originate from the violation of law by human agency; they can, then, certainly be prevented.

It is admitted by physiologists that all parts of the body can be changed by proper exercise and the law of nutrition,— some parts increased in size and strength more than others,— so that in this way a far greater measure of health can be secured. It is found that decided improvements can be made in the physical system during the lifetime of an individual, and that in three or four generations the human constitution may reach a higher state of perfection. If Nature has, therefore, established a physiological standard of health, — which is much less liable to disease,— and at the same time it is well understood that this standard is attainable, should not the greatest efforts be put forth to secure and maintain such a boon? It is by this means that the germs, the primary causes of a vast amount of disease, are to be forestalled. In this warfare with disease we have been content to lop off a few branches, leaving intact the trunk and roots. We have been battling the enemy in the outskirts, without attempting to take the citadel. Here is a great work for sanitary science; in this field it is destined to reap its richest harvests. It may take time; but reforms in which the highest welfare of mankind are involved never remain stationary.

In drawing this discussion to a close, a few suggestions may seem appropriate. While quoting from Doctor Flint's address on " Medicine of the Future," the inquiry arises, Is not sanitary science also to have a *"future"*? Most assuredly. Its past history is very brief and different from that of medicine. This extends back thousands of years, and its whole history is made up of a succession

8

of changes. It is not so with sanitary science. A half-century covers its whole existence. Its only change has been the constant unfolding and applying of Nature's laws to the improvement of health and prevention of disease. It has not been found necessary in its progress to try experiments or apply any new medicine. As sanitation is based upon the laws of Nature, its course can not change or go backward. Excelsior is its motto.

This sanitary movement has certain advantages over other reforms. Its success does not depend upon the medical profession alone, nor upon the patronage of government, nor upon any one body of men, but upon all classes,— men and women. The more people become enlightened on the subject, the more earnestly will they engage in the work, and become at once partakers in its benefits.

The history of sanitary science is full of promise for the future. It is really only about twenty-five years since it could be said to have had a fair start. Its doctrines have become deeply rooted, not only in the medical profession, but among large numbers of the laity scattered throughout this country and Europe. The press is committed decidedly in its favor. Its teachings are found broadcast in books, journals, pamphlets, reports, and newspapers. Its principles are being taught and applied both in our common schools and higher institutions of learning. Boards of health have been organized in all large cities and in nearly every one of the United States. The benefits already derived from this science can not be estimated in figures, or described in language. The pestilence in this country has been stayed ; epidemics have been checked ; a vast amount of sickness prevented, and a great multitude of lives saved. In Great Britain, where the science has made greater progress, and more exact accounts kept, upon Mr. Edwin Chadwick's authority, based on the registrar-general's report, it is asserted that the lives of 30,000 persons are annually saved, and 300,000 cases of sickness every year prevented by means of this science !

The two following statements, though once quoted, are so prophetic that they will bear repeating:—

Forty-two years ago Dr. Elisha Bartlett said, in Philadelphia, while urging upon the profession a more thorough knowledge of the causes and nature of disease : "The next thing to be done is to find out the best method of modifying and *preventing* disease. This is the great mission that lies immediately before us ; this is to constitute the great work of the next and succeeding generations." Forty years ago Dr. John Forbes, in an address to his brethren, said in London : "Redoubled attention should be directed to hygiene, public and private, with a view of *preventing* diseases on a large scale, and individually in our sphere of practice. Here the surest and most glorious triumphs of medicine are to be achieved."

Ten years ago Dr. Henry I. Bowditch, of Boston, who has given more thought to this subject than any other man in this country, said, near the close of his work on "Public Hygiene in America" : "We stand now at the very dawn of the grandest epoch yet seen in the progress of medicine. While philosophically, accurately, and with the most minute skill, studying by means of physiology, pathological anatomy, chemistry, the microscope, and above all, by careful clinical observation, the natural history of disease and the effects of remedies, — our art at the present day looks still higher, viz., to the *prevention* of as well as to the *cure* of disease."

These testimonials speak for themselves. They need no comment. The predictions here uttered are certain to be fulfilled. The glorious triumphs spoken of will surely be achieved. Individuals, organizations, and institutions may perish, but these principles will live and advance step by step, from one triumph to another, from one glory to another.

Prevention of Crime.*

A LITTLE more than one hundred years ago John Howard published his celebrated work on the prisons of England and Wales. From that period a new element entered into the treatment of prisoners. They began to be treated more like human beings, with the possibility of reformation. The principles of humanity and philanthropy, prior to that time, had scarcely been recognized in dealing with this class of persons. But during the first half century—from 1778 to 1830—the reform made very slow progress. About this latter period commenced two great experiments in prison discipline, denominated the "congregated system" and that of "solitary confinement." These experiments created much discussion, which gave a new impulse to the reform.

During the last half century, and especially within twenty-five years, many changes have been made. New questions in prison discipline have continually arisen, and experiments of various kinds have been tried. While some nations have shown more interest, and made, perhaps, farther advances in reform than others, each has had its own system of discipline with its peculiar merits. In our country important improvements have been made, but still many defects exist. We fail to reap the results which ought to be obtained, especially in the reformation of prisoners and the prevention of crime.

* An attendance upon the International Congress on Prisons in London, and many years' experience connected with the prisons of Massachusetts, have suggested this paper.

The leading objects of prison discipline may be expressed under these three heads : punishment, safety, and reform. As to the importance of securing these three objects, all, we believe, are agreed, however they may differ in the modes of doing it, or on minor points. These objects are here placed in the order in which each has come up in history, and in accordance with the prominence which many persons seem to attach to the whole subject; but, really, reformation should be the main object of confinement and punishment, for it is more important than both of the others.

The true test of such discipline is in the proportion of prisoners discharged, improved or reformed on the one hand, and on the other, of those released unimproved and perhaps made worse. If the former object were generally secured, there would be few re-committals and far less increase of criminals. But, unfortunately, statistics show that crime is constantly increasing, that prisons of all kinds are everywhere crowded, and that nearly one-fourth of their inmates are re-committals. It is very evident that there is something wrong in the present methods of dealing with criminals; that the ends attained do not correspond with the means employed; that *reformation* in the prisoner is not generally effected. Without attempting any thing like a full discussion of the subject, we propose to notice it briefly from a single point of view, viz., *the prevention of crime.*

There are two stand-points from whence facts and arguments may be brought to show *how* crime can be prevented : first, we may check crime in criminals ; and, second, keep persons from becoming criminals.

While, in the present state of society, it may be impossible by human means to prevent some persons from entering upon a criminal career, much may be done to break up such habits when once formed, and to produce a radical change of life. It is an encouraging fact, that wherever the proper means have been brought to bear, the character of prisoners has been improved, and the number of criminals reduced. Unfortunately this part of prison discipline has been altogether underrated or neglected, while confinement and punishment have received relatively too much attention. Reform in habits and character has been considered of minor consequence. In fact, such has been, not unfrequently, the treatment of criminals, both in spirit and manner, as to produce any thing but a reformatory influence upon them. There are laws of mind as well as of body, which, if violated, serve to make the individual more of a criminal. Let the animal and selfish nature in the prisoner be mainly exercised, without developing his moral and intellectual powers, and no reformation in his habits or character takes place or can be expected.

As this treatment of prisoners involves laws which lie at the foundation of all reform, and, of course, of correcting criminal habits, it should receive the most scrupulous attention. The criminal is a human being and governed by law. Crime is the violation of law, — not merely civil, but moral, — a law of Nature. A criminal has all the faculties of other persons, but not well-balanced or properly enlightened and trained. While the safety of the community and the principles of justice require that the criminal should be confined and punished, it should be done in harmony with *laws which develop his higher and better nature.* No criminal was ever reformed by being treated in a brutish manner, nor by appeals only to his animal and selfish nature. The conscience must be enlightened, and the intellect instructed. Hence, educational, moral, and religious influences ought to be brought into requisition far more than they are. No prisoner should ever be discharged without being improved, if possible, in his habits and character. This reform must be an individual work, the treatment varied or adapted to

every case. Reform seldom, if ever, occurs in classes or
large numbers at once. Such a work requires time, labor,
and means ; but its results will abundantly pay. It is in
this respect that prison discipline fails : its subject is not
reformed, neither is crime checked. It is in this direction
that improvements should be made, which would have a
powerful influence in the prevention of crime.

The most important reform in prison discipline that has
anywhere occurred is in Ireland. This is based upon a
system of improved classification of rewards and promo-
tions, as well as of encouragement and assistance to prisoners after being released. This reform commenced some
twenty-five years ago, and has resulted in a great decrease
of crime, as well as of the number of convicts. Improvements of a similar character have been introduced into the
English prisons, and have been attended with decidedly
beneficial results.

We regret that we can not report favorably in this respect of prison discipline in our own country. Since the
Civil war, crime has everywhere rapidly increased, and in
some states it is estimated that it has doubled. No thorough and systematic measures are taken to reform prisoners ; the old ranks are kept good, and new recruits are
constantly being added. Great pains are taken to provide
prisoners with labor, to make contracts, and to obtain as
much money as possible out of their work. If we had
the same amount of effort expended in reforming their
habits and character, crime and criminals would decrease.
There is great need of legislation on this particular point,
and that the managers of prisons have the power and be
specially instructed to adopt efficient measures for the
reform of prisoners.

The second means of preventing crime, viz., keeping
persons from becoming criminals, opens up a very important field of inquiry. It may be convenient to discuss

both these agencies that shape and mould human life after birth, and those that beget and fashion the constituent elements of the brain prior to birth.

All history proves that the criminal class, as a body, originates from a peculiar stratum or type in society,— sometimes from the middle or common walks of life, but more generally from the lowest orders, especially from the ignorant, the shiftless, the indolent, and dissipated. Occasionally we find some from high life, from those in good circumstances, from well-trained families, from good homes; but these are altogether the exceptions. In such cases any new or additional means to reform them would be useless. If our object, then, is to prevent crime on a large scale, we must direct attention to its main sources, —to the materials that make criminals; the springs must be dried up; the supplies must be cut off. This fruitful source or fountain-head of crime is found among children of poor or miserable parentage, surrounded by bad influences and exposed to all manner of temptation, with no one disposed or qualified to train and educate them. Such are the hot-beds of vice and crime. On account of indolent, shiftless, and dissipated habits, such people continue poor and dependent; large numbers also become orphans, and fall upon the public for support, or become pests to society. All such children should be picked up and gathered into reformatories, or find homes in good private families, which are the best reformatory institutions in the world.

Again: into such low communities, abounding with children under the influence of vicious parents, moral and spiritual education should be carried, and the state or municipality should see to it that every child is brought into the public school and educated. Efforts in various other ways should be used to bring moral and religious influences to bear upon such families and neighborhoods. Children become vicious and criminal frequently on account of their surroundings and immediate exposures. The sooner children can be removed from such localities and circumstances, the better. The public are not sufficiently awake to these upas-trees of poison, — these pests and hothouses of vice. All legal and moral means should be employed to reform such classes. Every state and city should adopt systematic measures for this purpose. The saving in taxation alone will pay for it. In no other way can crime be so effectually prevented.

More prolific than any other in the production of crime is the vice of intemperance. This operates in so many ways that it is impossible to trace out all its destructive effects. It impoverishes people, and brings them into circumstances of temptation; it corrupts the morals, and poisons the blood; it excites the evil propensities, and develops the animal nature ; it stupefies conscience, and destroys the moral sentiments ; it impairs in man the powers of free agency, and converts him into a brute. Whatever produces such effects upon the human system must have a powerful influence in the production of crime. The evidences come from all quarters, — and without contradiction from any, — that intemperance is the cause or occasion of three-fourths of all the crime committed, some estimating it even higher. The habit commences early and more readily with individuals and families who are predisposed to idleness and to low, animal life. The natural instincts of such persons flow in one direction, and drinking becomes a master-passion. If intoxicating drinks can be withheld from this class of persons, their habits and character become gradually improved. Total abstinence would do much to save them from a life of vice and crime. By this means more than half of the crime committed would be prevented.

But the primal and principal cause of crime exists in a

state of things prior to birth. That the "child resembles the parent," and "like begets like," are acknowledged truisms. It is also true that this resemblance or likeness extends to all parts of the brain, and, of course, to every faculty of the mind. If the lower and posterior part of the brain is predominant and continuously active, the animal propensities and selfish faculties will take the lead in character. If the parents are addicted to habits of dissipation and sensuality, the children will be predisposed to the same. If these habits are inveterate, the propensities are transmitted in an intensified form.

These transmitted qualities are more marked, and have a much wider range, than is generally considered. The blood itself may be tainted, and affect the structure and function of every organ in the body. Such may be the physical development as to incline one to lead an idle, low, and dissolute life, without ambition or self-respect. A living or means of support must be obtained without work or rendering an equivalent. There may be a strong will and an overmastering passion of selfishness, so that the individual is not inclined to be governed by the principles of justice, or to regard the rights of others. Such persons become an easy prey to temptation.

The celebrated "Margaret, mother of criminals," reported in New York a few years since, furnishes a striking illustration of hereditary crime. An investigation was made, through the New York Prison association, in the jails and prisons of the state, extending back six generations, which resulted in tracing out nearly three hundred criminals descended from one wicked woman! If a thorough inquiry were made on this subject, doubtless other similar illustrations would be found. If the truth could be known, we believe a large amount of crime would be traced back to hereditary influences.

How much crime might be prevented if certain classes

of vicious persons could be hindered from propagation ! What right have such individuals to bring upon the public so much misery, shame, and cost ? Within a few years laws have been passed for the forcible removal of nuisances and other evils, either injurious to health and life, or detrimental to the welfare of the community. With the rapid progress of sanitary science and the great advances in legislation, some means, we believe, will yet be devised for preventing, at least to some extent, the evils growing out of criminal heredity. Let the public mind be enlightened on this subject, let parental responsibility be placed on high ground, and the evils of improper marriages be pointed out, and within two or three generations a great amount of crime would in this way be prevented.

Changes in New England Population.*

IN the history of a nation or people there are sometimes important changes taking place so gradually and quietly that they are scarcely perceptible at the time. It may require a series of years or several generations to work out the problems involved, but they may be followed with results of great magnitude.

Some changes of this character have been taking place in our New England population, which we purpose here briefly to notice. In the earlier history of New England there were few changes in the residence of her people. As agricultural pursuits constituted their principal occupation, the same farms and lands continued to a great extent in the same families from generation to generation. Prior to the Revolutionary war very little emigration took place out of New England. In the early part of the present century many persons removed to New York, and some to Ohio. From 1810 to 1830 this emigration continued steadily to increase, not only to those states, but to the states and territories farther west. To such an extent had this emigration been carried on that, in 1840, the United States census reported nearly half a million of persons born in New England who were living in other states.

Whenever new lands were thrown into market by the government or by means of railroads, or some new mining

* This paper appeared in the *Popular Science Monthly* for August, 1883. It attracted much attention at the time, and has frequently been called for since, from all parts of the country.

interests, then a "Western fever" started up, and great
numbers might be seen "going West." While we have
no means of ascertaining the exact number removing from
New England, during any one year or period of time, the
United States census gives, every ten years, the birth-
place of all people residing in every state at the time the
count was made. The census of 1880 reports that the
whole number born in New England, but living in other
states and territories, was 566,848. This number is made
up by emigration from the different states as follows:
from Massachusetts, 175,349; from Vermont, 117,590;
from Connecticut, 108,797; from Maine, 93,256; from New
Hampshire, 49,397; and from Rhode Island, 22,459.

From another point of view it will be seen how these
natives of New England are distributed: New York has
133,272; Illinois, 53,128; California, 46,908; Iowa, 38,-
170; Michigan, 37,865; Wisconsin, 37,615; Minnesota,
34,636; Ohio, 32,819; Pennsylvania, 26,787; Kansas, 19,-
338; New Jersey, 18,148; and other states under 10,000,
and much less. Vermont has sent away the largest number for its population, and New Hampshire the least.
Maine and Massachusetts have sent the largest delegations
to California, being three-fourths of all the emigrants in
that state from New England. It appears by the census
that the states bordering on New York — Vermont, Massachusetts, and Connecticut — have sent over 100,000
persons to that state, while the other New England states
have sent only some 20,000. The representation from
New England (178,207) in the Middle states is much larger
than is generally supposed. This emigration has now
been going on for over three-fourths of a century, and it
would constitute a fact of great interest if we could ascertain the number of persons born in New England who
have ever removed from her borders to the Middle and
Western states, as well as to the territories.

The census of 1850 shows that at that time there were 454,626; in 1860, 562,997; in 1870, 615,747, and in 1880, 566,848.   It will be seen by these figures that for twenty years the number has been very stationary, the new emigrants making not quite good the number who had deceased.

It is full two generations since this emigration commenced.   As nearly all those persons emigrating were between the ages of twenty and forty, great numbers must have died at various periods.   The exact amount of this mortality it is impossible to ascertain, and the data for forming any thing like a correct estimate are altogether too uncertain ; it may have been a quarter of a million, and possibly a half million.   What has been the effect of this steady and large drain of people on New England, opens a question of much interest.

Without entering upon the discussion of the subject, we make two or three suggestions.   It will be admitted, we presume, that those young men and women leaving their homes possessed, as a general thing, more physical energy and mental stamina than those remaining behind. Such a loss of physical vigor and character must have had a decided effect upon business interests, as well as the present state of society.   But, from another point of view, the loss may have had a more decided and lasting influence, — that is, in its permanent effect upon physical and mental development.   The better the principles of physiology are understood, the more we discover what a powerful influence physical organization has upon the character of a people.   The permanent prosperity of any community depends far more upon the laws of inheritance than is generally supposed.   Let the most enterprising and promising among the young people emigrate from a place, and it must, in the course of time, have its influence. Whether the vital interests of New England have not suffered in this respect, from so many persons emigrating in the prime of life, presents a question worthy of careful consideration.

Interchange of Population. — There is another change going on in these states quite different from the one described. This consists of frequent removals from one state to another.

The census of 1880 shows that Massachusetts had at that time 68,226 residents born in Maine; 54,088 born in New Hampshire; 26,869 in Vermont; 20,514 in Connecticut; and 17,067 in Rhode Island, making 186,764 persons who have removed there from other states. At the same time these five other states had 85,478 persons living in their bounds born in Massachusetts. Deduct these 85,478 from the 186,764, and Massachusetts gains over 100,000, mostly from Maine, New Hampshire, and Vermont.

There is very little migration from the other New England states to Connecticut or Rhode Island, and scarcely any from the latter to the former. Massachusetts, Connecticut, and Rhode Island make very nearly equal exchanges, neither gaining nor losing much. These removals from one state to another are prompted from a great variety of interests, personal and local. The states most benefited by them are those employed largely in manufacturing business. These changes are carried on chiefly between villages and cities, and seldom take place in the rural or country districts. It may be said that the foreign element is largely concerned in these removals.

Country Life Exchanged for the City. — This change is not governed at all by state lines. It commenced forty or fifty years ago, from country districts to places where trade or business demanded help. The introduction of manufactures and mechanical pursuits of various kinds, as well as the opening of railroads, created a great demand for laborers. By means of those changes and other agencies, trade and commerce became very much enlarged, and furnished employment for increased numbers.

Here and there new centers of business were formed ; new villages sprang up, and large towns were converted into cities. In some parts of New England these removals have taken place to such an extent as to change the face of the country and the state of society. It commenced first in the small farming towns, and has prevailed most in places remote from markets and railroad accommodations.

The effect of such removals is especially marked in Massachusetts, as she possesses a larger number of cities, more railroad facilities, and a greater diversity of pursuits. The census shows the following facts : That of 345 towns in Massachusetts, from 1845 to 1855, there was a decrease of population in 86 towns; that from 1855 to 1865 there was a loss in 166 towns ; from 1865 to 1875 there was a loss in 142 towns, and the census of 1880 reports a loss in 143 towns.

It will be seen that the number of towns losing population varies at each census, but undoubtedly the same towns are reported as decreasing in numbers each decade. It should be stated that, in about one-quarter of those towns, the loss was occasioned by a division of the town or annexing a part of it to some other place. It should also be stated that the removals from the country districts to villages and cities do not account for all these losses of population ; emigration to the West, and to other distant places, does a part of the work, and so also does death.

There is another item in the account : the birth-rate has so much declined in rural districts, that scarcely any addition, if any, comes from *natural increase*. But, as the death-rate in many places exceeded the birth-rate, the thinning out of the people is not confined to Massachusetts.

In Maine, New Hampshire, and Vermont, the hill towns and many of the agricultural districts are losing more or less population — not alone by death or emigration of young people, but by the removal of whole families to more populous places.   In Rhode Island and Connecticut there is not the same extent of territory, and population is more equally distributed ; but still the census of Connecticut reports a decrease of population in some sixty towns in the western part of the state.   Statistics show that this removal of people from the country to the city has been increasing every year ; and when it will cease, or what is to be the result, time only can tell.

Agriculture as Related to Other Pursuits. — Connected with this decrease of population in country districts, there is one very important consideration : that it involves a change of occupation.   Farming is given up for work in the store, the shop, and the mill.   Within half a century the business of New England has passed through great changes.

By the censuses of 1860, 1870, and 1880, we find instead of an increased number engaged in agriculture with the increase of population, that the number has been actually diminishing.   The census divides all kinds of business or occupation into four classes :   1, Agriculture ; 2, Professional and personal service ; 3, Trades and transportation ; and, 4, Manufactures and mechanics.   An examination of the tables representing these four classes in the reports of 1870 and 1880 shows that the last three classes have increased relatively far more than the first class.

The number engaged in agriculture has fallen off in every state.   Vermont and Massachusetts stand in respect to agriculture at extreme points ; the former has more people engaged in farming than in all other pursuits, while the latter has only about one-tenth as many employed on the farm as are engaged in other pursuits.

Maine has the largest number of any state engaged in agriculture — about one-third of her whole population —

9

and she at the same time possesses the greatest amount of territory to cultivate. New Hampshire has half as many engaged in agriculture as in all other occupations ; Connecticut has one-fourth, and Rhode Island only one-tenth. The whole number in New England engaged in agriculture was 301,765, and in other pursuits, 1,268,116 — more than four times as many. In 1870 the proportion was one to three.

A comparison of this table (1880) with that in the census of 1870 shows a far greater increase in the class of professional persons than in that of any other occupation or pursuit. The census of 1870 reports only 145,324, while the census of 1880 reports 349,984 persons. This increase is found in every state, though in some states greater than in others. Whether this great increase of professional persons in ten years is an indication of an improved state of society or not, is a question upon which there might be differences of opinion.

It is well understood that fifty years ago farming constituted the principal occupation of New England ; but, instead of maintaining its position, with a greatly increased population, it has fallen far behind other pursuits. The great additions made to her people have been absorbed in trade, in manufactures, and mechanical business. In considering this exchange of agriculture for other pursuits, a question of great interest arises : What is to be its effect upon physical organization and the permanent prosperity of a people ?

No fact is more firmly established than that agricultural pursuits are the most healthy of all, and that those engaged in them transmit *physical development* in its best estate. All experience proves that an exclusive city population tends gradually to degenerate physically, and that the stock can not be kept good from generation to generation.

It is well understood that the only conservative power that can prevent this degeneracy in cities, is that their population shall constantly be replenished by recruits from the country. But it should be borne in mind that the places in the country made vacant by these removals are soon occupied by a different race of people, and that this foreign element is pretty likely to increase more and more in the farming districts of New England.

Supposing this change should generally take place in the country districts, how is the purely American stock to improve or be kept good? It can be done only by an intermingling of the races, which is even questionable.

Change in Birth-rate. — There is no agency so closely connected with the vital interests of a people as the matter of the birth-rate. In the history of nations this has always been considered a question of the utmost importance. To a certain extent, it operates as a thermometer to show the rise and fall of national prosperity. The process of its operations may seem slow, but certain results are sure to follow.

In respect to this agency a most surprising change has gradually been taking place in New England. Near the close of the last century, Malthus, after making a survey of all the nations on the earth, selected the United States (virtually New England, which was the most populous part) upon which to base his theory of population. Seeing that the inhabitants of these states doubled in twenty-five years by natural increase, he considered that it afforded most favorable indications of prosperity. At that time the birth-rate was high, families were large, and few were found without children.

From the first settlement at Plymouth, in 1620, this prosperous state of increase continued without much change for nearly two hundred years; but early in the present century some decline in the birth-rate commenced.

It is impossible to trace the exact changes which have taken place for the last two or three generations.

In some parts of New England the precincts and towns were accustomed to keep very correct records of all births, but they were not generally printed, so no comparison of them can be made; but for thirty years or more several of the New England states have published registration reports of births in their cities and towns, so that very correct comparisons can be instituted. Without going into a detailed sketch, by statistics, figures, etc., of the changes in birth-rate, we present some general statements on this subject. Forty or fifty years ago large families, numbering six, eight, ten, and twelve, were quite common; now they are rare, — in fact, a large number of such families can not at the present time be found in any one neighborhood, or even in a single country town. Formerly, in the rural districts of New England, there were few families having only one, two, or three children; and in case there were none, it was so rare as to attract particular attention, and was considered by many a great calamity. But what a contrast is found in the present state of society! In the great majority of our American families only one, two, or three children are now found, and in very many families not *one*. And such a state of society is approved by the fashions and prevailing sentiment of the day!

As registration reports generally return the births of the foreign population in the same tables with the American, and as the term *native* is applied to all infants whose parents were born in this country, though of foreign descent, it will be at once seen how difficult it is to obtain the exact birth-rate separate of each class. Two facts are pretty well established: 1, That the birth-rate of the foreign class is more than twice as large as the strictly American; and, 2, That, in the country districts of New England, settled mainly by the Americans, it is question-

able whether the birth-rate exceeds the death-rate, — that
is, there is no addition to the population by natural increase.
Should this birth-rate continue to decrease as it has for
the last twenty or thirty years, the effect will become more
and more manifest than it has in the past. The board of
health for New Hampshire, having charge of the registry
of births and deaths in the state, in their report just published, state an important fact bearing on this point.
After carefully analyzing the births and deaths in 1880,
to draw the line between the foreign and the American,
the board make out that the deaths among the Americans
exceed the births by eight hundred, — that is, New Hampshire lost population from this source. If this same test
of birth- and death-rate as reported in New Hampshire
should be found to apply to all the other New England
states, the record would not be very creditable for the
past, nor encouraging for the future.

In making comparison between birth- and death-rate,
the latter must always be carefully taken into account.
If the death-rate is unusually large it affects at once the
gain by natural increase. In New England the death-rate
generally is not high, which is more favorable for the rate
of increase. The same is true in Great Britain, but the
birth-rate is much higher there than here. Thus large
additions are made there to population by natural increase;
far more than in New England. In France for several
years the death-rate has been rather high, so that allowance
must be made. As a matter of fact, the comparison with
foreign nations is decidedly unfavorable to the New
Englander.

According to the latest and most authentic reports, the
birth-rate of the New England states is less than that of
any large European nation, except France. And this
birth-rate of New England is based upon both the foreign
and American classes; could the latter be eliminated from

the former, it would make the birth-rate of the strictly American even much lower than that of France. It is well understood that the population is steadily decreasing in certain portions of France, and that this decrease is every year extending. This decline in numbers is attracting more and more the thoughtful attention of the French savants, and the inquiry is made for the causes and the remedies. It may be found to resemble certain diseases, the causes of which can readily be discovered, but the remedies can not easily be applied.

Foreign Population in New England. — Of all the changes in New England, the introduction of the foreign element is the most important. The facts respecting the history of this immigration, and the extent to which it has reached, can be obtained, but no human sagacity can fully foresee its results. There are, however, certain features in these changes which should be carefully studied, and the developments or tendencies growing out of them should be better understood. More facts, more knowledge, are needed on this subject. What, then, is the history of this movement? Fifty years ago the foreign element in New England was very small. In Massachusetts the census reports that in 1830 it was only 9,620, and increased as follows : in 1840 it was 34,818; 1850, 164,448; 1860, 260,114; 1870, 357,319; and 1880, 443,402. It should be borne in mind that these figures represent only the "foreign-born," and not their children or descendants, which would greatly increase the number.

In the other New England states the whole foreign element combined is not so large as that in Massachusetts, and has not increased so fast. In Maine, in 1850, it numbered 31,450, and in 1880 it was 58,883; in New Hampshire, in 1850, 13,571, and in 1880, 46,294; in Vermont, 1850, 32,931, and 1880, 40,959; in Rhode Island, 1850, 23,111, and in 1880, 73,993; and in Connecticut, 1850,

37,473, and in 1880, 129,992. The whole number of foreign-born in New England, reported by the census of
1880, was 793,122, and 360,649 of these emigrated from
Ireland.

The census reports the whole population of New England, born in the United States, as 3,234,317; but large
numbers reported here as *natives* are of foreign descent.
It is impossible here to draw the line, but, from the best
evidences before us, we should say there must be about
half as many in this class as that of the foreign-born,
which would increase the foreign element to 1,200,000 in
New England.   It may be larger.   The *Catholic Directory*,
six years ago, stated that there were at that time 890,000
souls in New England connected with that church, and
the number must have since considerably increased.
Then, of the 793,122 reported by the census "foreign-
born," there must be a large number of Protestants, —
being over 100,000 emigrants from England and Scotland.
The same organ also six years ago stated that "nearly
twenty-five per cent. of the population of New England
is composed of Roman Catholics." The census reports
the whole population of New England as 4,027,439 in
1880.   At the present time (1883) the foreign element
must number over 1,200,000 persons in New England.
But it is quite unequally distributed.   In Massachusetts,
Connecticut, and Rhode Island it numbers more than a
third of the population; but in Maine, New Hampshire,
and Vermont it is not one-quarter.   As the birth-rate of
this class is more than twice as large as the American,
the foreign element will constantly gain in numbers upon
the American.

Connected with this large addition to our population,
composed of a different people in race, type, and character, there are several points that deserve careful consideration.   A few years ago it was thought that emigration

from Ireland would very much diminish, if not cease; but of late it has taken a new start, and may again flourish. Emigration from England and Scotland is sure to continue; so also from the British Provinces and Canada. But this foreign element is destined to increase hereafter more by births than by immigration. The marriage-rate is much higher in this class than the American. It is possible that, in the process of time, changes in the style of living, and by adopting modern fashions, the birth-rate of this class may be somewhat reduced, but certainly not at present.

Religious influences have a powerful hold upon this class of people, so that they may be restrained from violating the laws of the physical system. In process of time there may be such a change in the organization of this people as to reduce the birth-rate. The *Catholic World* stated six years ago that "nearly seventy per cent. of the births in New England were those in Catholic families." This estimate we thought at the time was too large; but with the increase of births since belonging to this class, and the addition of the births of large numbers of the foreign-born and foreign descent who are not Catholic, it will increase this percentage.

In most of the cities more than half of the births for years have been connected with the foreign element, but it was not expected that the same proportion could be found to exist in rural districts and country towns.

It does not seem possible that three-fourths of all the births in New England at the present time can be classed under a foreign head, but the indications are pretty certain that such will be the case before many years, and then we shall be compelled to believe the fact. The inquiry is frequently made, If the two classes do not intermarry, what is the prospect in this direction? There are occasional intermarriages between the American, the English, the Scotch, and the emigrants from the Provinces, but not often between the Americans and the Irish. Still, cases of this kind do occur occasionally between the laboring classes, and we think they are increasing. The registration reports divide certain married parties into two classes, — the foreign-born father and native mother, and *vice versa.*

The term *native* here might apply to the strictly American, but a careful examination shows that each party called native was of foreign element, so that there was no mixing of the two races. This class of marriages has been constantly increasing. In Massachusetts, according to the registration report of 1881, there were 7,386 births of this class ; nearly one-eighth of the whole number.

Change in Physical Organization. — The most serious evil resulting from the introduction of this foreign element is in causing a change in the physical organization of New Englanders. In the case of men, that part of farming requiring hard work, those kinds of mechanical pursuits demanding physical strength, and, in fact, nearly all manual labor out-of-doors, have already passed mainly into the hands of foreign help. This change, commencing thirty or forty years ago, has everywhere been taking place, but more rapidly of late years.

This exchange of regular physical exercise for lighter employment and in-door work is calculated to develop nerve tissue, rather than the muscles ; to impair the power of digestion, and reduce the vital forces of the system. That a course of physical degeneracy to some extent has thus been going on with New England men, must now, upon thorough examination, be generally admitted.

But a change, more marked and serious in its character, has been taking place in female organization. Formerly all kinds of house-work and domestic duties were performed by New England women. Before foreign help could be obtained, our young women were generally employed as domestics in families. It was customary for the more wealthy and many families of the middling class, where there were no daughters, to employ one or more domestics. In many families all the house-work was done by the daughters and mother, without any imported help. It was considered becoming and praiseworthy for women, both the young and middle-aged, to engage or hire in domestic service.

All such employment was then considered respectable. Skill, fidelity, and success in domestic duties were the best recommendations that any young woman could possess. Practice and public sentiment in these respects have entirely changed. Very few girls of pure American stock can now be persuaded to engage in domestic labor. Such service is generally considered by them menial, and every kind of employment or business away from the kitchen and domestic hearth is preferred. In families where there are daughters, the hardest portion of the house-work is now performed by the mother or hired help.

What are some of the effects of this change in domestic life? No kind of exercise is so well calculated to develop all parts of the body in the female, and promote good health, as house-work. No study or employment can fit the young girl so well for house-keeping as practical training in such duties.

In this way home and the family are pretty sure to secure a strong attachment. By these means all parts of the body are harmoniously developed; a sound constitution, good health, and long life are secured. Instead of educating the girl in accordance with the *laws of her physical system*, and training her for the great practical duties of the family, from the age of ten to eighteen she is kept at school nearly all the time, so that the brain and nerves are developed at the expense of other organs. This partial and one-sided development of the body is increased and intensified in the female, by being thrown out of her natural sphere in domestic labor and family relation. Hence, great multitudes of young women, from fifteen to twenty-five, have nothing to do, are everywhere seeking employment, and are constantly exposed to an excited or morbid state of feeling.

The ill-health of New England women is proverbial. It is less than half a century since it attracted public attention. A careful examination will show that its history and extent run almost parallel with the high pressure in education and the neglect of house-work. The nerves and the brain have been cultivated at the expense of the muscles and physical stamina.

In this artificial state of society wants multiply, and fashion has a powerful influence. A high and extravagant standard of living is set up, and young people are unwilling to commence life as their fathers or grandfathers did before them. For twenty or thirty years there has been a steady decline in the marriage-rate. There are powerful influences, starting partly from internal sources and partly from external agencies, which threaten the permanency and best interests of the family. If the laws of the human system can be so changed or violated as to defeat its primary objects, this institution must suffer and decay. There is a normal and healthy organization of the body, as well as of the brain, which favors married life and the family relations. On the other hand there is such a thing as an abnormal development of the body and a morbid condition of the nervous system, which is decidedly unfavorable to the domestic relations; especially is this the case with females.

The law of maternity is already violated to such an extent that it is questionable whether half our New England women can properly nurse their offspring. There is a general law in nature that "supply and demand" go together and are co-equal, and if one fails, the other is endangered. There are also decided evidences that the maternal instinct, — love of offspring, — one of the strongest and holiest instincts of our nature, is fading away.

It should be borne in mind, that when the harmony or balance of organization in the body is materially changed, — that is, certain parts obtain an extreme development, while the functions of others become very much weakened,—a similar change and derangement of action appear in the brain. The fact is well established, that certain portions of the brain perform distinct and separate functions. Let that portion of the brain, whose functions pertain to the family relation and to domestic life, fail in proper development and healthy action, and supreme attention be given to the culture of the intellect and moral sentiment, and, in process of time, its effects on character will become very manifest. If this change in mental development applied only to an individual here and there, its effects on society would not be so marked or injurious ; but, when the great majority of persons are affected by it, the results become far more extensive and serious in their character.

There is a higher, or rather a normal, standard of physiology, by which all changes in physical organization can be tested, whether it is improving or degenerating. This is based upon the perfect structure of every organ and the legitimate performance of all their functions in a normal, healthy manner. The nearer human organization approaches this standard, the greater is its power or ability to secure the highest objects of life. The great excellency of this standard is its balance of power between the body and mind; its harmony of functions as developed by all the organs of the body, as well as every part of the brain.

Again: the family constitutes the foundation or ground-work of all society; and, when properly established, is the most powerful agency in the world for human improvement. This institution must have its basis and supplies in the social and domestic affections, guided by the intellect and controlled by the moral sentiments. Without such a foundation it can not be made permanent, happy, and prosperous. The intellectual faculties will never alone cement and perpetuate this institution.

Some singular developments on this subject have recently been brought to public notice, —that is, in matters connected with the subject of divorce. Among no other civilized people is there such a breaking up of the family. Why should it occur here, among a people so highly educated and moral? Some attribute it to changes in legislation; but the primary causes of the evil existed before, and will continue, in spite of any changes in legis-lation. Its outward developments may by this means be checked, but the evil is not cured. The primary causes of these anomalous developments have, we believe, to some extent, a broad and deep foundation in *physical organiza-tion*. We do not see how all the facts connected with this alarming evil can be accounted for in any other way.

There is one consideration connected with this whole subject, of vast importance, which can here only be men-tioned, —that is, *heredity*. The changes in organization are directly and most intimately connected with hereditary influences. The effects of such changes through these laws are so great and far-reaching that they can not be described or measured.

Hereditary Influences.

THE fact of hereditary influence was early observed. It was proclaimed in the times of Moses. Numerous illustrations of it are found in the Sacred Scriptures. It was taught by the Greeks and Romans, as well as by many able writers since their day. But it was not till near the close of the Eighteenth century that systematic attempts were made to improve in this way the stock of domestic animals.

By careful study and close observation it was found that experiments in this direction proved very successful. Great changes, both in Great Britain and in this country, have been made in improving the qualities and character of domestic animals. To such an extent have these experiments been carried that they have been reduced almost to a regular science. The same general principles that have been employed in the animal creation apply also to the human species.

Physiology, upon which these principles are based, is comparatively a new science. Within a few years great progress has been made in the practical application of this science, and just in proportion as we study the relations existing between the parent and the child, or between one generation and another, do we find marked indications of hereditary influences. It may be safely said, we believe, from the numerous testimonials and illustrations on this subject, that there must be much truth in these laws of inheritance.

But a great difficulty or barrier stands in the way of progress on this subject,—that is, the want of a *general principle or law*, by means of which all the facts or knowledge of this kind can be classified and reduced to a system. In all departments of natural history, or sciences in their early stages, there is a period of experiment, of observation, and discovery before the facts can be classified and arranged under general principles. It is this kind of work—the establishment of a general law—that the facts of heredity need more than any thing else, and such is the object of this paper.

In all the works of Nature, primary laws or general principles are perfect in their character, for they are based upon an ideal of perfection. This rule holds good in all the natural sciences. Thus in tracing back hereditary influences to their primary source or origin, the presumption is, that they stand upon some ideal standard of human perfection.

After many years of observation and reflection, we venture to submit a general law or standard, upon which all hereditary influences are based, and from which they have their origin. This law, of course, is based upon physiology. What, then, is this ideal standard? *It consists in perfectionism of structure and harmony of function.* Now, let one or more of the organs become changed in structure, and impaired in discharging its proper functions, the effect, more or less in degree, is transmitted to the offspring.

It will be seen at once how weaknesses and predispositions to disease may be transmitted. Suppose there is an enlargement of the heart or some valvular difficulty, or suppose the lungs may be weak or some part of them diseased, the effects of such an organization are quite likely to be transmitted in this direction. The same law governing the body applies also to the brain. If certain portions of the brain are imperfectly or excessively developed, thereby causing weak or strong points in the character, similar developments and characteristics will be found in the child.

Let us illustrate this law by taking some striking facts in heredity, such as appear in the defective classes — the idiotic, the feeble-minded, the blind, the deaf-and-dumb, etc. The law is based upon a normal, healthy standard of the whole body — every organ normal in structure and performing its natural functions in a healthy manner. This presupposes that the brain is well developed and performing its legitimate work, and, also, that the senses of sight and hearing are sound and healthy. Now, would such an organization beget offspring idiotic, feeble-minded, deprived of sight and hearing? Assuredly not; it would be impossible. While we do not find perfect organizations, but only approximations to them, yet the nearer we approach them, the less such defects are likely to occur. Suppose this physical standard, naturally sound and healthy, has become impaired, — some parts abused and diseased, — then these imperfections will be transmitted. This law of hereditary influence applies to the brain and to the senses, as well as to all other parts of the body.

It has long been admitted by the best writers on medicine that there is a large class of diseases called hereditary, from the fact that the germs or predispositions to these complaints are transmitted. There may be instances where the disease can not be traced back to the parent or grandparent, but may have existed in some of the ancestors, passing over one or two generations. The diseases most likely to be transmitted are consumption, scrofula, rheumatism, neuralgia, disease of the heart, liver, etc.

Perhaps there is no organ in the body where the predisposing causes to disease are greater than in the brain. It is estimated that fully one-third of all the insanity may be traced directly, or indirectly, to hereditary influences. The brain, from its delicate structure and incessant activity, is more likely to be disturbed, and its functions become more or less deranged, than almost any other organ in the body. If the morbid, diseased action of one organ implies that there must be a normal, healthy standard, why may not all these be combined and make a general, universal standard? And why should this not constitute a general law of heredity, from which all minor points have their start and origin?

If we could always have the same data, — the same organization upon which to base hereditary influences, — the results would be determined more definitely. But in applying this law of heredity we encounter a serious difficulty at once: there must be two active agents, not possessing the same organization, which may be constituted widely different. It is in this union, or combination of similar and dissimilar qualities, that the results or effects of inheritance must be estimated. As a general thing, where there is great similarity in the agents, there will be sameness in results; while on the other hand, the greater the differences, the more widely marked the results.

One of the most important elements in constituting a good organization is that there should be a balance or harmony in the organization. In this case we shall not find marked excesses or defects; and provided both parties possess such an organization, it is almost certain that the offspring will have sound and healthy constitutions. The same principle applies to the brain; if its parts are not well developed, — some excessive, and others deficient, — the mental qualities of the child will not be evenly balanced.

In entering into matrimony it is desirable that the parties coming together combine such organizations as

10

complement each other; those qualities wherein one is deficient, the other should make up. This conduces greatly, not only to the interests and happiness of the parties themselves, but it insures favorable hereditary results. In order to secure such advantages there is need of understanding this general law of heredity.

In making application of the law, it presupposes that other conditions are favorable; such as the age, the union, and the adaptation of the married parties. Provided no natural laws are violated or interfered with, there will uniformly be found with such an organization, not only the greatest number of children, but they will be endowed with the greatest amount of physical vigor, strength, and health. It should also be added that with such an organization the best development of all parts of the brain might be expected, giving balance and symmetry to mental qualities, whether social, intellectual, or moral; in fact, it is the highest and most perfect development and standard which Nature sets before us.

This organization consists briefly in the perfectionism of structure and function; or, in other words, is the normal standard of anatomy and physiology in their highest and best estate. Upon this basis is founded not only the law of human increase, but also the general law of health and longevity. Weaknesses and diseases originate in deviations from this standard, or in violation of some of its laws. Thus, in the changes taking place in the human body there are general principles to guide us, and a universal standard of appeal. By this means clearer views and more definite knowledge can be obtained of all weaknesses and diseases to which the human body is subject.

There is a great advantage in having a standard of organization constantly before the mind, as it enables us to detect more readily, in every case, what diseases are constitutional or hereditary. We can thus judge far better of the relations which one disease sustains to another. This knowledge will also enable us, not only to treat this class of diseases more successfully, but to understand how they may be prevented.

Now, a careful investigation will show that it is the constitution or organization here described that survives the longest or reaches the greatest age. It is this type of the physical system or combination of forces that insures longevity; and the most powerful of these forces is that of heredity. All writers agree that one of the indispensable requisitions for long life is good healthy stock, or long-lived ancestry. If there is any condition, property, or principle that composes or regulates these inherent qualities and tendencies, there must be some general rule overruling the whole.

There is another test in favor of this normal type of physiology ; that is, it is the true standard of beauty. In the creation of man there must have been a standard, a form, a size, a fullness, a proportion, an outline, etc., that was more beautiful than all others. Man was created with a sense of taste; with a love for the beautiful, which, cultivated and perfected to its highest state, might find objects in Nature capable of gratifying it to the greatest possible extent.

The physical standard here described represents the organization of man as perfect, — the same model and type that it was when he came from the hands of the Creator. It is this same standard or model that Grecian and Roman artists have attempted to imitate in statuary. Art may create such models, but what a failure on the part of Nature ! What countless deviations from this standard do we find everywhere, among all people ! What has been the most powerful agency in producing these changes? It is the *law of inheritance*, first and foremost,

above all other agencies. Why should not such a power
be better understood? Why should it not be more under
the control of the human will?

Within a few years the interest in this subject has
greatly increased, as indicated by the publication of several new works, as well as by discussions in the journals
and newspapers. Some advocates of the doctrine are so
enthusiastic that they claim, if the principles of heredity
could be generally applied, it would improve the present state of society; that it would go far to eradicate evil
and crime, as well as pauperism and insanity. In their
zeal for this new doctrine they overestimate altogether its
advantages, and do not consder the difficulties in the way,
or how slow must be the process of improvement. It is
the work of successive generations.

On account of the advocates of the doctrine making
such high pretensions of what it can do, some persons
have become very much prejudiced against it, and ridicule
its followers. It is not the first or the only time that new
doctrines have been opposed and ridiculed. This arises
in a great measure from ignorance and prejudice. The
facts on this subject are so common and abundant that
they must convince every candid and reflecting person
that there is much truth in them.

The principal reason why the laws of inheritance have
not hitherto been better or more generally understood is
because the principles of physiology have not been applied any more to practical life; in fact, this science is
practically in its infancy. It is only a few years since the
relations between pure air and the healthy state of the
lungs and the blood became known, or the importance of
regular exercise of all parts of the body, in order to maintain good health.

The relations which the physical system, with its various
organs, sustains to education and religious culture, are, as

yet, very imperfectly understood. So is the application of sanitary laws to public health; also to the prevention of disease and the preservation of human life. The farther inquiries are pushed into the relation which this science sustains to the public welfare, the more useful and important do they appear. It may be found that this law of inheritance will become one of the most powerful agencies that can be employed for advancing the best interests of a people. Such an agency certainly should not be despised or ignored.

The inquiry may still be made: If the doctrines here advocated are so important, why have they not before become generally known and their truth admitted? The same question might have been raised in reference to many other discoveries. It seems to have been the design of Providence that the great truths of Nature should slowly be brought to light, at different periods, and by a variety of agencies. Such has been the history of nearly all the sciences. A great amount of knowledge may exist on some subjects without being reduced to sytem or applied under general principles.

It is so in regard to heredity. A large body of facts have been gathered by a great number of individuals, each operating in different fields. Now let all these facts be carefully analyzed and classified, to see if some general principles can not be deduced from them,— some principles which will enable us to understand better their origin, their connection, and application. In the very nature of things there must be some general law to explain and regulate these phenomena.

In review of the facts here stated, we ask if they do not afford sufficient data and argument to claim some attention? Let the reader take the *normal standard* of physiology as here described and study it carefully from all points of view; let him select individuals and families among his acquaintances, and see what are the deviations in their case from this standard; let him examine into the relations between parent and child, and see what are the resemblances, what physical and mental qualities are inherited. The more striking and peculiar the organization and character of these persons are, the greater and more marked will be the hereditary effects. Let him take the defective classes, such as the feeble-minded, the blind, the deaf-and-dumb, and the insane; let him select cases from the highest and lowest grades in society, and examine into the character of the offspring,—physical and mental,—and we are sure he will be convinced that there is such a thing as the law of inheritance; and, if it is true in one single case, there must be a great universal law covering the whole.

The New England Family.*

W HY the *New England* family the subject of this
paper? Because it is a historic family; it is the
root and seminal principle of American civilization. The
ideas born and nurtured in it are permeating this whole
nation. Fifty millions of people have received a social,
political, moral, spiritual impulse from it, and the end is
not yet.

The peculiarities of such a family deserve to be better
understood. Also, if there are any agencies threatening
the stability and best interests of the family in this highly
favored portion of the land, they should be exposed. It
is evident that some changes, anomalous and difficult to
explain, are taking place in the population of New England.

As the family is the natural source of increase in population as well as the fountain head of social and national
prosperity, a careful study into the influences that affect
this institution, and the changes that are being wrought in
it can not but be profitable.

It may be well to notice, first, certain principles in
physiology which have a direct bearing upon the subject.
This science is comparatively of modern growth, and may
justly be said to be in its infancy, as it respects some important applications.

* This paper appeared in the *New Englander*, March, 1882, and many
calls have since been made for it.

Once physiology was studied chiefly with reference to disease and individual health, — its relations to public institutions and the welfare of society generally not being well understood. Within forty or fifty years the relations of physiology to education have received much attention, and more recently, special interest has been awakened in respect to its bearings upon hygiene. The more thoroughly this science is investigated in its varied relations to human welfare, the more extensive will be found its applications, as well as valuable its results. But in respect to the family and the laws of human increase, the principles of this science have not been investigated or applied as they should be. It requires no argument to prove that physiological laws have a most direct and intimate bearing upon both subjects. While the laws of population and the family institution have been discussed from different points, the laws that govern the human body in multiplying the species have not as yet been properly investigated and explained. Here physiology is destined, we believe, to achieve its noblest triumphs.

The most distinguished writer on population, T. R. Malthus, makes very little account of this science. In all his discussions touching changes in population, its increase and decrease, there is no allusion to the operation of the laws which govern the human organization. The science of physiology was scarcely known in his day. The most important English writers succeeding Malthus — M. T. Sadler and T. Doubledey — laid more stress upon physical organization, but failed to establish any general principle. The course pursued by Herbert Spencer and Charles Darwin is very different. Yet while they had made important discoveries in physiology, and discussed its application in a variety of ways, the laws that govern population and the family institution have received from them very little consideration. We venture to submit here a general law or principle on this subject. At this time we can present only the substance of this law, referring the reader to papers in which it is more fully discussed by the writer.*

This law is based upon a normal or perfect standard of the human system — where every organ in the body is complete in structure, and performs all its natural functions. This implies that the body is symmetrically and well developed in all its parts, so that each organ performs its proper function in harmony with the others.

While this perfect physiological standard may only rarely be found, there are approximations to it in great numbers, some much nearer than others. The fact that the law is based upon physiological organization does not preclude the influence of other agents, such as climate, food, government, etc. ; but these factors are secondary. Also, in carrying out this law, it is presupposed that other conditions, such as age, sympathetic union, and mutual adaptation of the married parties, are favorable.

With such an organization there will uniformly be found — provided no laws of Nature are violated — not only the greatest number of children, but they will be endowed with the highest amount of physical strength, health, and mental capacity. On the other hand let the body be developed to extremes in either direction, towards a predominance of nerve tissue with a large active brain,

* "The Laws of Human Increase." *Quarterly Journal of Psychological Medicine*, April, 1868. D. Appleton & Co., New York.

"The Physiological Laws of Human Increase," vol. 21. Transactions of the American Medical Association, 1870. Philadelphia.

"Lessons on Population"; suggested by Grecian and Roman history. *Congregational Quarterly*, October, 1871. Boston.

Lecture on Hereditary Influences before Massachusetts Board of Agriculture. Transactions, 1872. Boston.

"The Normal Standard of Woman for Propagation." *American Journal of Obstetrics*, vol. 9, April, 1876. Wood & Co., New York.

or towards a predominance of the lymphatic muscular temperament ; either extreme will be found very unproductive of well organized children. This normal standard of physical development applies also to the brain. It is of the highest importance that all parts of this organ should be well developed and their functions harmoniously performed.

This is especially necessary in respect to the marriage relation. Whatever differences of opinion may exist among physiologists as to the functions of the brain, on minor points, it is generally agreed that the social and domestic affections have their seat in the lower posterior portions, while the moral and religious depend upon the upper portions, and the intellectual upon the frontal lobes. To ensure the perpetuity and the best interests of the family, all the parts or all the faculties of the brain must co-operate.

Connected with those views two important considerations should be borne in mind: 1st. That exercise increases the parts, physical or mental, which are used, while neglect of exercise diminishes them. Thus a constant change for the better or the worse may be going on in the organization and character of an individual.

2d. The establishment of a general law of population affords an explanation of the laws of inheritance. These spring from the former, and constitute one general plan ; for without a general law or guiding principle these hereditary influences can not correctly be understood, or successfully applied. It is through the brain that mental qualities are transmitted ; and by the adoption of a normal standard of physiology, applied both to the body and the brain, the laws of inheritance become intelligible. The physiological laws which have been considered in respect to population, hold an intimate relation to the family. The more thoroughly this science is investigated, the more extensive and practical shall we find its principles as applicable to every-day life. It will enable us better to understand individual peculiarities, and the relations we sustain one to another, especially in domestic life.

This will appear more and more evident in the discussion of the subject now before us.

The value and permanency of the family as an institution can not be too highly estimated. It is indispensable to all civilized society.

It is the nursery of the church, and no people or nation can prosper long without it. Wherever in the history of the world a people have attained the greatest prosperity, or advanced to the highest civilization, there the interests of the family have been most sacredly guarded and preserved. It was so in the prosperous days of Greece and Rome. The same facts were true in regard to the Jewish nation.

In Great Britain, from its earliest history to the present time, the family has been looked upon with a sacred reverence. Civil authority has surrounded it with the strongest safeguards, and the church has regarded it as one of divine appointment. On nothing have the affections of the English people centered more than on the home, and nowhere else have the relations of the family been more permanent than in the land of our fathers. The greater the culture and refinement, and the stronger the religious element, the purer and more sacred has been the tie that has bound families together.

Such is this institution in Great Britain at the present day, and such was its character among the Puritans in the early history of New England. The family stood with the church in the respect and affections of the early settlers. The relation of husband and wife, of parents and children, was held as most sacred. The mutual interest of the parties in these relations generally lasted through life, their sympathies growing stronger and stronger, and their attachments more tender. For two hundred years the homes of New England were well nigh models. The families were generally large, and lived in a kind of patriarchal style. The government centered in the united head of the family, was usually administered in kindness and in accordance with high Biblical principles. Seldom was there a separation between the husband and wife, or was there discord among the members of a family. The instances of breaking away from parental authority were rare, and it was not often that a son or daughter turned out badly in life. The fruits of good discipline, faithful instruction, and early training in the family, were everywhere visible.

In the history of no people, probably, can there be found better illustrations of well-ordered families than for two hundred years were found in New England. While the religious training was prominent, the education of the intellectual was carefully attended to, and the social and domestic affections were most wisely developed.

But the New England family is not the *same* now that it was two hundred years ago. Changes that deserve careful consideration are taking place with respect to its character and permanence. That we may better understand these changes, and the dangers that threaten the family, let us inquire briefly what is its true foundation — what its primary objects. The family has a two-fold foundation — the sacred Scriptures and a man's physical organization. The Bible teaches that this institution was established in Eden, and all through the Old Testament and the New, it is regarded as the corner-stone of the church and state, — in fact it was itself the early church and state. The family in all places in the Bible is treated as of divine appointment. The family institution is based also upon physical laws which are a counterpart of Revelation. These laws always harmonize perfectly with the revealed will of God, when both are correctly interpreted. Independently therefore of the Scriptures, or of any divine teachings, or examples, we believe the necessity of such an institution as the family can be proved from physiology alone.

There are fundamental laws in this science which clearly point in that direction, and can not otherwise be correctly understood or properly observed. Moreover, it may be proved that the race can not be perpetuated in its best estate or highest development without such an institution.

Taking this view of the subject, we shall find the applications and observance of these laws to be of the greatest consequence.

What then are the teachings of Revelation and Science as to the primary objects of marriage or the family? The teachings of the former are briefly stated or summarized in works on the subject, in formulas of marriage, in liturgies and prayer books, by both the Protestant and Catholic churches throughout Christendom, as follows: 1st, The production of children. 2d, The preservation of chastity; and 3d, Mutual company and help.

That these simple propositions constitute the primary objects of marriage can be substantiated by the testimony of writers of the highest authority connected with all the leading religious denominations, both in this country and in Europe. The experience of family life in Christian nations, where there has been the greatest amount of culture, morality, and piety, confirms also the truth of these statements. It may be added that these propositions are not antiquated or superseded; that whatever religious changes have taken place, or modern fashions have been adopted by society, the objects and relations of the family remain the same. Many changes have been made in society which may be considered real improvements, but it is not so with the family; its laws are fixed and unchangeable. But it is not left for the sacred Scriptures alone, nor even experience, to prove the truth of the propositions here stated. The laws of the physical system afford the strongest possible evidence.

Such is the construction of the organs of the human body, including the brain, and such are their functions, that if exercised normally, they are adapted to secure all of the objects here specified. If there is a real defect in the structure or function in any part of the system, or if there is a failure in securing either of the objects stated, it goes so far to weaken or impair the marriage relation. It matters not whether this occurs through ignorance or design, these laws can not be violated with impunity. If this violation of law is designed with a motive prompting to it, and a will sactioning it, the injury is not mere physical, but a most serious one to the mind, and destructive of moral principle. Let such violation be often repeated and long continued, and the injury to the body and the whole spiritual nature is incalculable. No one but a physician can fully appreciate it, or forecast its consequences.

Nowhere have these primary objects of marriage been secured in a higher degree than among the earlier settlers of New England. Nowhere has the marriage relation been happier, more permanent, and attended with better or greater results. But has there not been within fifty years a wide departure from the examples and teachings of these settlers? Has there not been a marked deterioration in the sacredness of this relation, and have not the influences and motives leading to the formation of marriage sunk to a lower plane? In short, have not the primary objects of marriage been more or less lost sight of, and the relation been considered very much in the light of mere partnership, intended to promote simply the convenience and self-interests of the parties? If a change of this character has been made from what may be considered a normal, healthy, religious basis, to one abnormal, artificial, and selfish, it is very evident that its fruits or results must prove unfavorable.

The family as an institution must be based mainly upon the domestic affections, guided by the intellect and controlled by the moral sentiments; otherwise the relations can not be either happy or permanent.

But a change in this respect has been gradually taking place among a large class in New England. The cause of the change is a greater diversity in pursuits, modes of doing business, the powerful influence of fashion, a higher style of living, and a more artificial state of society.

The general introduction of foreign help, we believe, has had in a variety of ways an injurious effect upon the family. It has caused that all kinds of domestic service — extending to that performed by the members of one's own family — is looked upon generally as menial and degrading. Once, American girls in large numbers were employed to do house-work, which not only educated them in the best possible manner to perform all such duties, but at the same time improved their constitutions and gave them physical strength.

No kind of work or exercise is so well calculated to develop all parts of woman's physical system as house-work. The variety of it, being always at hand, and suiting itself to one's convenience, would seem to indicate that Nature expressly designed it for the healthy development of woman's constitution. In this way not only good health and a knowledge of domestic duties are obtained, but also the best fitness and qualifications possible for the relations of the family.

Few American girls can be found at the present day who are willing to engage in domestic service, and those living at home do only the lighter kinds of work, — the harder portions being performed by their mothers or hired help. By neglecting work in the kitchen and the home, and seeking pursuits that tax chiefly the brain and nervous system, the young woman not only changes her habits and character, but her organization. Large numbers from five to fifteen years of age are confined in school, and have very little time for work or exercise. Thus the brain and nervous temperament of the young woman become unduly developed, and she fails to secure that physical strength and stamina which are indispensable for discharging the duties and responsibilities of the family.

As a result of this change of physical organization and supreme devotion to brain development, pursuits and objects of interest are sought away from home. Thus a desire is created for a more artificial life, a higher style of living, and hence a multiplication of wants.

To such an extent have these things gone, that there is set before the young New England people a higher and more expensive standard of living than the majority have the physical strength or pecuniary means to support. Young men contemplating marriage and finding they can not adopt this standard of living, are compelled to postpone it till they can obtain the means, or they give it up entirely.

When matrimony is seriously contemplated, its delay usually multiplies these obstacles, and the plan fails of completion. Those who postpone it are exposed to many temptations, and often find that the habits formed in single life become fixed, so that when they do marry, the relation does not prove what they expected, and instead of a happy or perfect union, unpleasantness and discord mar their life. Also, those who enter upon married life early encounter many difficulties. They must live in boarding-houses, or at hotels, or have rooms in one place and take their meals at another. In case they commence housekeeping, finding they can not live in just the style they wish and have their wants gratified, they resort to more rigid economy and closer calculations in expenses. The questions of personal comfort, of fashion, and self-interest have a controlling influence. For preventing or reducing expenses, a sharp eye is had to those things that cost the most, rather than to what is most useful in making a virtuous, healthful, and happy home. With too many young women the fashions of the day and the attractions of society have more influence than the duties and enjoyments of home. The monotony and confinement of domestic life become irksome, its cares and labors burdensome, so that no increase in numbers can be encouraged. The least change in that direction is looked upon as so much additional care, burden, and expense. By this and other means, the foundation of marriage is transferred from the domestic affections to the selfish sentiments and the intellect. What must be the effect of such notions on the conduct and character of individuals? Do they not tend to make them more selfish and weaken the domestic ties? While these remarks may seem to reflect more particularly on women, they are not so intended; man is a party and partner in all this.

Again: What is the effect of this change in respect to the primary object of marriage, viz: the production of offspring? This is no trivial question. Language can not express its importance in its bearing upon the family. All we can do here is to state a few facts and inferences.

For half a century or more the birth-rate in New England has been steadily diminishing, and for the last twenty or thirty years much more rapidly than at any former period. Our birth-rate is now lower than that of any European nation except France, and when confined to the American class, it is lower than that of France.

The birth-rate and the death-rate are approximating in New England, so that it is very questionable whether in

11

many places there is any increase of population among
native New Englanders. The increase is confined almost
wholly to the foreign element. The birth-rate among the
Irish, Scotch, English, and Germans is twice as large as
among the Americans.

Among the foreigners are many large families, and only
a few married couples without children; whereas among
Americans there are many married people who have no
children, and very few that have large families.

What a contrast do the families of the present day present to those of one hundred years ago, in which were
eight, ten, twelve, or more children? In 1875, a census of
Massachusetts was taken by families, which brought out
some striking facts. While the number of foreign families is much the smaller, they report a large majority of
the births. In tables representing the number of children
at different ages in American and foreign families, there
is a surprising contrast. In the former, there are only a
few large families, and many consisting of only one, two,
or three persons. The census reports at that time in Massachusetts 359,009 families; of these, 23,739 consisted of
only one person; 115,456 of only two, and 140,974 of
only three persons.

While the census does not report the number of children in these three classes — making in the aggregate
279,569 families — it is evident there were very few. The
great body of the children must belong to 79,446 families,
constituting the balance. A large proportion of this class
of families is undoubtedly foreign. As to the increase of
population reported by the census and other authorities,
we are in danger of being misled. If no foreign element
had ever settled in New England, or if this class could
now be entirely eliminated, the changes in population
would be far better understood. It would be seen at once
what the increase is and whence it came. It would de-

velop a class of facts, we apprehend, that for a civilized
people has no parallel in history.

There is another way in which both census and registration reports may deceive us. These documents base
their tables or figures upon nativity, and not upon nationality. Some registration reports have endeavored to distinguish between the two classes, so that reliable data
might be furnished, showing the relative growth; but finding so many difficulties in the way, it is now generally
abandoned.

Thus the transition from foreign to American is rapidly
taking place in New England. All born here are called
"Americans." It should be borne in mind that the census,
in reporting the largest class of families in Massachusetts
as consisting of but three persons, includes the foreign
element in the count. If this enumeration of the family
in the state had been confined to the settlers, it would
make quite a difference. It is unfortunate that the real
facts in the case can not be obtained. Few persons are
aware how rapidly this foreign element is increasing in
New England. In several of the states the annual registration reports have returned for years a majority of births
as being of this class. The school children in nearly all
the cities are composed largely of those of foreign descent.
The *Catholic World*, published in New York, estimated,
in 1877 that nearly seventy per cent. of all the births in
New England belonged to Roman Catholic families. That
estimate seemed large at the time, but it is undoubtedly
too small now. Since then over one hundred thousand
Canadian - French have come into New England, who
abound in children. It is estimated that foreigners compose but a little more than one-quarter of the whole
population.

There is another test of the family institution which
indicates a deterioration. Within twenty or thirty years

there has been an alarming increase of divorces in New
England. These are confined almost entirely to Protestants and native Americans. The statistics on this subject have been carefully collected and tabulated for years.
It appears by these tables that divorces have been steadily
increasing in all the New England states. The latest
return (1879) is as follows:—

One divorce in Connecticut to eight marriages; one to
nine in Rhode Island; one to thirteen in Vermont, and
one to fifteen in Massachusetts. As Maine and New
Hampshire have never published any official reports of
divorces and marriages, the exact ratio for these states
can not be given; but from partial statistics obtained, and
from other sources, it is very evident that they make no
better showing than Vermont and Massachusetts. The
records of the courts show also that about one-fourth of
those applying for divorce fail to obtain it, on account of
adverse evidence or opposing facts developed in process
of trial. From the large number of divorces and the exposure it makes of personal and private matters, the presumption is that there must be many more families where
discord and variance exist, but they decline to bring their
troubles before the public. It should be stated that most
divorces are obtained within a few years after marriage,
and generally there are few or no children, even if the
parties have been married many years. It should also be
stated that among no other people or nation, at the present
time, do we find divorces to such an extent; and at no
former period in history have they ever been so numerous,
except in Greece and Rome, shortly before their downfall,
and in France a little after the French revolution.

Why should there be in New England so great a prevalence of divorces? It might be supposed, from the social,
educational, and religious influences existing, this would
be the last place where the family, the most important of

all our institutions, would thus be broken down. It does not arise from any alteration in the laws, for the demand came first, and as the appeals became more urgent and numerous, legislation has been changed or modified in their favor. Divorces have greatly increased in several of the Western states, where society is constantly changing, and things generally are in a more unsettled condition. It might be expected that there marriages would be hastily formed and on too slight an acquaintance; but in the state of Ohio we find a singular development of facts. As these have a most significant bearing on several points under discussion, we quote the following statement from a lecture given Feb. 25, 1881, by Rev. S. W. Dike, in the Monday course of lectures at Tremont Temple, in Boston : —

"In the Western Reserve, comprising the twelve northeastern counties of Ohio, settled mainly by emigrants who went from Connecticut long before that state entered upon its new departure in divorce, and containing, it is said, a purer New England stock than can be found in the entire country, unless it be in parts of Maine, the ratio of divorces to marriages was from 1 to 11.8 for the two years, 1878 and 1879, while in the rest of the state it is 1 to 19.9. Nor is the worst of the Reserve in the cities. The ratio in Ashtabula county, among a farming people originally from New England, is 1 to 8.5 ; and in Lake county the proportion of divorce suits begun to marriages is 1 to 6.2, and of divorces granted, 1 to 7.4. Unless there be like counties in Maine, this is the worst county in divorces in the United States, except Tolland county, Conn., as that was for a few years. But if you go down to Gallia county, peopled with Welshmen and Southerners, the ratio is 1 to 50, and in Coshocton, 1 to 47.2. The divorce-rate in these counties of the Reserve is several times what it is in these and other counties. I am told, too, that the birthrate in Ohio is lowest where the divorce-rate is highest. It is said that the people of these counties are the most intelligent and virtuous in the country, and that the law-abiding citizens of the Reserve go to the courts for divorce, while those in other counties do not."

In the same state and under the same laws, why should there be this difference in the ratio of divorces? Why should there be one divorce to every *six* marriages among the "most intelligent and virtuous" people in Ohio, while there is only one in *fifty* among the less cultivated and refined portions of the same state? It should be borne in mind that the people occupying that part of Ohio designated "Western Reserve" are in their origin New Englanders, transplanted to a richer soil and a more widely extended territory. They are understood to have New England religious principles, educational advantages, and domestic habits; but these people are generally composed of the second, third, and fourth generations of those once emigrating from Connecticut. After so many years, and so long a separation, why should the family there partake of the same character as in Connecticut, in respect to the fewness of children? May not the reasons be the same as in New England, — a departure from the primary object of marriage, but especially a radical change in woman's physical organization?

No satisfactory explanation of these facts about divorce can be given except as we refer them to a physical cause. Changes in the laws, differences in the religious character, in educational advantages, and family training, can not fully account for them.

The second object of marriage is chastity. Many things show that marriage now fails to secure this object to the extent it should. That adultery is brought forward so prominently among the charges against one of the married parties, to obtain divorce, is pretty sure proof. In Massachusetts this charge is proven in more than one-third of the divorce cases; in Rhode Island and Connecticut about one-third, and in Vermont one-fourth, making for the whole fully one-third. In the attempts to obtain divorce, it is well understood that adultery is a difficult charge to substantiate in court, so that if the crime is actually proved in one-third of all the cases, there is reason to believe that licentiousness prevails extensively, not only in married life, but far beyond those bounds.

While it may be difficult to show to what extent or in what way marriage at the present day fails to secure chastity, there are some practices bearing on the subject which are alarming. That licentiousness is actually increasing in New England, it may be difficult to prove by facts and figures, but that covert vice of this character, in certain forms, is positively increasing, would be the testimony, we believe, of the most competent judges, — especially of members of the medical profession.

In order to obtain evidence on this subject, the Rev. S. W. Dike, of Vermont, instituted lately the following experiment, which we give in his own language. In his lecture on "Divorce," given at Boston, Jan. 24, 1881, Mr. Dike says: "I sent a list of questions to a hundred or more gentlemen in nearly all parts of New England, — mostly judges, state's attorneys, lawyers, police officers, large numbers of physicians, and specialists, with a few clergymen. Nearly all responded. About seventy letters are of value for the purposes of classification. These cover probably one hundred towns and cities, giving the opinions of nearly two hundred persons who were consulted in their preparation. I form, so far as these letters go, the opinion that there is probably less of open and coarse vice of certain kinds in many respectable country towns than there was seventy or eighty years ago; very likely less than there was fifty years ago. But with this exception, which covers but a part of our country towns and occasionally a city, as correct a summary of opinions

as I could give, would be like this : in three-fourths of
the localities reporting on this point licentiousness is said
to be increasing.   In nearly as many the destruction of
unborn life goes on as fast, or faster, than ever.   Physicians are very emphatic on this point, and many speak
with great indignation of the wicked practices of some
church members.   In one-half the places licentiousness
and drinking are found together, while one-quarter report
more licentious than intemperate persons in their com-
munications.   Nearly all find this increase among the
native population."

Few persons are aware how extensively this "destruction
of unborn life " is carried on, even in what are considered
the better classes of society.   But the "arts of prevention,"
which are also being extensively employed, are a far more
dangerous foe, not only to the family, but to the virtue
and purity of the community.   They open in a covert
way the flood-gates of iniquity.   If violations of law are
encouraged in married life, and found to be safely practised
there, the same things will be attempted outside, and the
primary object of marriage will be defeated.   Methods
that have long been employed in France have become not
only well understood here, but improved upon by Yankee
skill and ingenuity.   Besides the viciousness and crimi-
nality of such practices, their evil effects upon the physical
organization are incalculable.   Every physiologist who has
given the subject careful consideration will testify to the
truth of this statement.   The very existence of the family
is imperiled, we believe, more by these practices than by
all other agencies combined.

In the present state of society there is another way in
which licentiousness is increased.   In consequence of the
reduction of the marriage-rate and postponement of the
time of marriage, we now find a much larger number of
persons of both sexes, at a marriageable age, in the com-

munity than formerly. This increases the temptations to licentiousness, both within and without the marriage circle.

The fact has been well established by experience, that early marriages are the best safeguard to virtue and chastity. When we consider, in this state of society, how the practices referred to above are thought to secure the violators of law from exposure, what powerful temptations are presented to the passions! Virtue and moral principle do not afford sufficiently strong barriers to restrain excited passions, which, under such circumstances, can be so easily gratified. The eating of forbidden fruit creates a distaste for that which is pure and holy. The high and noble objects of the family institution are defeated; its true pleasures and enjoyments are lost, and mankind sinks to a level far beneath that occupied by the brutes. Chastity violated carries with it a dreadful penalty. From this Nemesis no guilty one, man or woman, can escape.

As to the third object of marriage, — "mutual help and company," — it is difficult to ascertain exactly to what extent this is secured. On account of the changes in the times and the different modes of doing business, men are kept away from their homes more than formerly. Such also are the competition in trade and the expensive style of living, that men are compelled to throw their whole energies into business, early and late, to the neglect of domestic duties. Such is the strain, the "wear and tear," upon body and mind, that men in active business break down early, especially in cities. Thus the rate of mortality has considerably increased among this class, and there is also a greater number of widows than formerly. Also men and women are drawn away from their homes more now than formerly, to summer and even winter resorts, for society, rest, and recreation; and these places are often scenes of temptation. There are also lodges, clubs, con-

certs, and saloons, which draw together the people.   The
attraction in either case is stronger than the attractions
of home.   The influence that centers in these places is
not always the highest, the taste cultivated not such as
ministers to the wants of man's nobler nature.   Many hus-
bands spend much time in this way, which they could more
profitably devote to their families.   Much of their hard-
earned money is thus unwisely consumed.   What they
give to gild and decorate public places would make their
homes a paradise.

While the introduction of foreign help into domestic
service has its advantages, it is at the same time attended
with many disadvantages.   Its effect upon the family institution deserves far more consideration than it has
received.   One obvious effect of employing foreign domestic help has been to impair the constitution and health of
New England women.   This is a most serious injury in
its relations to the family, and especially in its bearings
upon the welfare of the race.   It tends to prevent that
training and education of the American girl in the kitchen
and home duties, which are so essential in domestic life.
Instead of learning in the kitchen the art of good cooking, the girl now finds it more convenient to study it in
books, and to be taught it by lectures in school.   But
this mode of instruction frequently comes too late in life;
besides, very few even pursue this mode.   Good house-
keeping is a great home attraction.   Economy, neatness,
order, and good cooking are indispensable requisites to
the health and happiness of a family.

In no way are the bonds of the family so firmly ce-
mented and strengthened as by the comfort and happiness
of the home; and these depend much upon the persons
by whom household duties are performed.   It may be said
that the present mode of educating girls, and giving them
the advantages of greater refinement and higher accom-

plishments, makes them better company and more fit to take charge of domestic affairs. But if this course of education impairs their strength and injures their health, how can such girls become practical housekeepers? Invalids make poor housewives. In the marriage relation a sound constitution and good health are of primary importance. The lack of these is one of the greatest abatements to domestic happiness, and furnishes a prolific source of trouble in married life; and, when their importance becomes better understood, multitudes will be deterred by this alone from entering the marriage relation.

There are some general considerations that have an important bearing upon this subject. Registration reports show that the marriage-rate has been, for many years, lessening in New England, and also that marriages are more frequently postponed to a late period in life. The census of 1870 reported that fifty-four per cent. of the adult population in Massachusetts were unmarried, and only about one-half of this per cent. was under twenty years of age. This fact shows that a large number of adults in this state were leading a single life. If the foreign element could be eliminated from society, this proportion would relatively be much larger. The same fact applies to other New England states. The proportion of adults leading a single life is likely to increase. Such a state of society is abnormal, and operates in a variety of ways unfavorably to the family.

The causes that have led to this can in part be explained, but the remedies for the evil can not so easily be applied. The elements, physical and mental, that constitute the family relations and cement and perpetuate them, are not fixed quantities. They may become weakened or strengthened by our own acts; even radical changes in disposition and character may be effected by hereditary influences, and thus be transmitted from one generation to another.

Not only the family, but the highest interests of the race, are involved in these hereditary influences.

Formerly New England women found little difficulty in nursing their offspring; this is true at the present day with the English, the Irish, and the German women living in our country. But not more than one-half of our young New England mothers can now properly nurse their offspring, and this number is every year decreasing.

It is said that this neglect arises from unwillingness on the part of mothers. This may be true occasionally, but it is not the rule. Inability, from lack of sufficient development of the mammary glands, and the requisite power in the digestive organs, is the real cause. The nervous system has been developed unduly, while the sympathetic, affectional, and muscular natures have been stinted. The failure to nurse offspring shows something radically wrong. The like of it, to such an extent at least, can not be found anywhere, either in history or among the women of any other race or nation at the present day. Evidently the divine office of maternity — woman's bright crown, her grandest privilege — is here passing away. The laws of Nature make supply and demand co-extensive and interdependent. If mothers nurse their children, they have and keep the power; if they neglect it, they lose the power. If they have lost the power, some important law has been violated. There may be instances of natural infirmity, disease, defect, or idiosyncrasy of organization, which account for the inability to nurse. Such cases are very rare, and they excite no alarm; but when the inability becomes common, including perhaps a majority of married women, and is rapidly increasing, it may well awaken anxiety. If this goes on, what will the end be? What will the woman of the next generation, and her successors, become?

Connected with the inability of mothers to nurse their

children, another change of vital significance is going on, — there is dying out of our women the "love of offspring." This love constitutes the noblest and purest of all the instincts or affections in women. It was wisely designed by the Creator that this should become, especially in females, a leading feature in their character. Accordingly we find among the women of all tribes and nations this "love of offspring" very predominant; and the more enlightened and Christian a people are, the purer and stronger should be this affection. While this instinct is naturally one of the strongest, and is intended to have a dominant influence in life, the whole order of a normal, healthy state of society encourages and develops it more and more.

What could be more unfortunate than that we should find certain influences operating in society to stifle, suppress, and crush out this natural affection! The agencies so pernicious are not described in books, nor are they taught in schools, and very little talk is heard in society respecting them. Still they are all-powerful; as much so as self-interest and fashion can possibly make them.

There seems to be a settled determination among many New England women approaching or entering the marriage relation, not to be troubled with the incumbrances of children, or at least to make their number very limited. It is true this sentiment or purpose, so unnatural, so unbecoming, is not proclaimed publicly, but it is well understood by the parties themselves, and it is the general sentiment of what is called cultivated and refined society. They compliment those who have none or a very small number of children, while comments, not pleasant or becoming to hear, are made respecting those who have large families or a goodly number of children. This, we know, is a grave charge, and may shock the sensibilities of some and be repelled by others; but we think it is true. Connected with the above is another most baneful sentiment, which is gradually creeping into popular favor among young people, — that the bearing and rearing of children belong to low life and are degrading. Whenever married couples in city or country town are blessed with several children, remarks or insinuations are frequently made, reflecting upon them, implying that their life is vulgar and sensual. The manner and tone are more significant than the language. But the sentiment is not confined to verbal expression. It crops out often in a variety of ways in the popular magazines and books of the day.

How different the teachings and examples set before us in the Scriptures! How different the instructions and illustrations found in Grecian and Roman history! How different from the spirit and practice of the best society in Great Britain! And what a change in this respect between the women of the present day and those of a hundred years ago !

The "arts of destruction and prevention of human life," referred to in another part of this paper, are comparatively unknown among the Irish, English, and Germans of New England. But if the "arts" here practised, with all their ruinous effects, could be brought to light, they would make a terrible revelation. If physicians should tell all they know on this subject, it would make a shocking disclosure.

There is one place in particular where the maternal relation is brought to the test, and to which physicians are not unfrequently witnesses. We can not describe it better than in the language of one of the oldest and most distinguished physicians in Boston, who has had a very large obstetric practice.

In the December number of the Boston *Medical and Surgical Journal* for 1879, Doctor A—— says : "In the early part of my practice the prevailing fashion and desire among married women were to bear children and rear families. They possessed the courage, and were willing to suffer for such a boon. To be barren was considered among the Jews a curse of the Almighty, and many of our grandmothers cherished sentiments akin to this. *Tempora mutantur!* What physician at the present day has not had to hang his head for shame, and feel the strength of his moral indignation rise, at witnessing the apathy or positive dislike—to use no stronger term—with which the first faint cry of the new-born infant is received?—I can not say welcomed—by the friends, and perhaps by the suffering mother! . . . I have never known an Irish mother, no matter how poor, or how many little ragged children around her, that did not receive every new-born babe with emotions and expressions of gratitude, as a blessed gift from God. This sentiment, however rudely expressed, has never failed to win my admiration; and I take pleasure in pointing it out as the finest trait of Irish female character."

What a contrast do these two pictures present! How tender and natural the latter! how cold and heartless the former! If such is the reception of the infant at birth, what will be its care and treatment in life? Then, if the infant is deprived of its natural nourishment at the breast, and is fed artificially, there being at the same time a lack of strong natural affection for children, it is not strange that infant mortality has greatly increased. It is the testimony of physicians that no one thing is so conducive to the health and life of the infant as a plenty of good breast-milk, and this is generally accompanied with strong natural affection.

Again: If the love of offspring is constantly suppressed, and in the course of time in a great measure eradicated, what is to be its effect upon female character? Will it not destroy in a measure the family element in women?

Nothing strengthens the ties between husband and wife so much as children; nothing binds together a family so closely, and makes home so attractive, as the paternal relations. But if the primary object of marriage is defeated, soon the family will be a thing of the past.

In conclusion, we remark that the foes described as threatening the New England family have not come suddenly into existence, nor do they rest upon the surface of society. There is a lack of that patriotism which leads one to endure pain and practise self-denial to people this land. There is a wide-spread spirit of luxury, which makes both men and women seek supremely their present and personal enjoyment. There is too little of that high moral principle that prompts people to forget self and find their life in giving life and happiness to others. There is also too little value set upon the worth of an immortal soul. If parents can bring into existence and rear up for endless bliss a never-dying spirit, how great the privilege! But these considerations, so powerful in the early settlers of New England, have greatly ceased to influence our people. Selfishness, the love of ease and present enjoyment, and living for this world only, have become dominant; and they have already wrought so much evil that a change in woman's physical organization is one of the results. And this effect now itself operates as a cause to hasten on the ruin which is impending over the family.

Again: These foes are not the product of a true civilization, or of pure Christianity. They arise from a direct violation of the spirit of the latter and the laws of the former. The most effective means of reform consists in exposing the origin of these evils and their dangerous tendencies. The question may be asked, Have not these evils already got such headway that they can not be arrested? Certainly not. The family is not what it once was in New England. The difference does not consist in numbers merely, but a failure in the unity and strength of all the relations which make the institution stable and prosperous. Considering the great importance of the family in its influence upon society and human welfare, its purity and strength should be most sacredly guarded, and its welfare most earnestly sought. It may seem to some persons that the changes and dangers here described as going on in the family are overrated. Such may be the impression of those who have never given the subject much thought or consideration, and who look only upon the surface of society; but no one, who has carefully watched for years the undercurrents of influence, and at the same time recognized the powerful agency of physical laws in the formation of habit and character, can be of that opinion. It is the change of woman's organization, instinct, and character, — and that, too, in the wrong direction, — which clothes the subject with so much importance. But it is not the women, or family of the present, that alone suffer: it is the *type* of organization, the hereditary tendencies, that must be taken into account. These will be transmitted in an intensified form. The interests of the race and of generations are involved in the issue. It is *this* which gives the subject transcendent importance.

12

The Prevention of Insanity.*

INSANITY is to be ranked among the greatest misfortunes that can befall a human being. It may deprive him of his rights as a citizen, his right to manage his own property, his right to testify as witness in court, and the right even of his own person. He becomes at once an object of distrust and notoriety. He is liable to be forced away from his home and his friends, and be confined in a lunatic hospital. He is also subjected to great expenses, which his own estate must pay, or his friends; if these fail, the place of his residence, or the commonwealth, must maintain him. These expenses in time become very large, frequently consuming all the means of the insane and that of their friends, to such an extent that they have to be supported by the public,—that is, they become paupers. In case the disease assumes a chronic form it generally continues through life. According to established life tables, it is found that a man twenty years old, becoming insane, will have an average insanity of a little over twenty-one years. The least cost of supporting an insane person in a lunatic hospital can not be less than two hundred dollars per annum. Thus it will be seen that in twenty-one years the expenses of support will exceed four thousand dollars. But this is not all; there should be reckoned in the account the value of his services or earnings, which

* The thoughts in this paper were suggested by serving as commissioner of lunacy in 1874, and also by frequent visits to the lunatic hospitals while a member of the Board of State Charities.

would accrue to his family or to his friends, in case he were early restored to health. This would amount in the same time to a much larger sum.

Now, if this insanity could be prevented, what an advantage it would prove, even in a pecuniary point of view! The physical suffering, the distress of friends, the anguish of mind, the loss of reason attending a life thus spent, no language can describe.

I. *What is Insanity?*—In the whole history of medicine there is no disease about which there have been such absurd and contradictory notions. These can be traced back to the earliest period, both of profane and sacred history. Reference is made in the Scriptures, in several instances, to persons "mad," acting a "false character," and governed by some "evil spirit." A notion prevailed that in all such cases some demoniac or satanic agency had taken possession of such persons, and that surely they were not in their right mind. These views respecting insanity were generally entertained by the Jews, and, with some slight modifications, prevailed among the Greeks and Romans.

It is a singular fact, that in all those cases where "madness" was attributed to persons, it was believed they were afflicted or controlled by an "evil spirit"; that it was not from a voluntary internal movement, but that they were taken possession of by some secret, mysterious agency outside, which was evil in its design and foreboded no good. Such were the views of insanity entertained not only in those times, but which continued to have a powerful influence for centuries later. And notwithstanding their absurdity and extravagance, multitudes at the present day are more or less affected by these strange notions.

While the evidences of this "madness" were manifested through the body, singular views also prevailed with reference to the soul or mind—that it was an essence or entity, acting independently of the body, and governed by no fixed laws or principles. The theories entertained in respect to the mind were so vague and indefinite that they served to mystify the subject of insanity.

II. *Functions of the Brain.* — Early in the present century special inquiries were made in respect to the functions of the brain; and, after many experiments and extended observations, it was generally conceded that the brain, in some sense, was the organ of the mind. If, therefore, normal, healthy operations of the mind depended upon the brain, should this organ become disturbed or in any way affected, it would at once change the character and action of the mind. This theory enables us to understand much better what is meant by insanity, or mental derangement; and just in proportion as the various developments of insanity were studied in connection with the functions of the nervous system and the brain, the better they were understood, and the more rational and correct views prevailed. In this way comparisons could be instituted between insanity and other diseases. If the various diseases of the body arose from violating the laws that govern the healthy action of the system, thereby causing an unhealthy, abnormal state of certain organs, — then, if the healthy action of the brain were disturbed or affected, it might, on the same principle, cause mental derangement.

The more closely mental phenomena are studied, as connected with physical organization and laws, the better will they be understood. Within half a century great advances have been made in a better knowledge of the functions of the brain and the laws of disease ; and these lead to more rational and correct views of insanity. That the brain is the organ of the mind, seems now to be generally admitted. When, therefore, this organ is in a normal, healthy state, the operations of the mind are sane and rational ; but when the brain assumes a morbid and

unhealthy condition, the mind is disturbed; its manifestations become unnatural and deranged. The first disturbance of the brain may be very slight; so the first symptoms of mental derangement may be.

III. *Prevention of Insanity.*—In all the discussions on
insanity found in reports, journals, and books, there is
scarcely a reference to prevention till within a few years.
The most decided statement that we have seen appeared
in the seventeenth annual report of the Commissioners
in Lunacy for Scotland. This is so much to the point
that we are induced to make the following quotations : —

"It is impossible to come to any other opinion than that
insanity is to a large extent a preventable malady; and it
appears to us that it is in the direction of preventing its
occurrence, and not through the creation of institutions
for its treatment, that any sensible diminution can be effected in its amount. Lunacy is always attended with
some bodily defect or disorder, of which it may be regarded
as one of the expressions or symptoms.

"We must therefore attempt to prevent its occurrence
in the same way as we attempt to prevent the occurrence
of what are called ordinary bodily diseases; and if it be
admitted that to a large extent preventable diseases exist among us in consequence of this ignorance of the people, it is clear that we can only convert the preventable
into the prevented, by the removal of that ignorance
through a sounder education. Men must be taught that it
is their duty, and not merely their interest, to understand
the laws of health, and to make them eventually the rule
of their conduct. In short, we can only hope that preventable insanity, like other preventable diseases, will be
diminished in amount when the education of men is so conducted as to render them both intelligent and dutiful
guardians of their own physical, intellectual, and moral
health."

No higher testimony on this subject could be quoted
than that from the Lunacy Commissioners of Scotland.
Several distinct points are here brought out: 1st, That

insanity is a disease, and can be prevented as other diseases are; 2d, For this purpose similar means must be used to those employed to prevent diseases generally; 3d, The public must be better educated and trained in respect to the laws of health; and 4th, By this process only can we expect a diminution of the disease. Lunatic hospitals alone will never do it.

IV. *Sanitary Science.*— What, then, has been done to prevent disease ? No regular or systematic movement of this kind was made till some twenty-five or thirty years ago. From careful investigations into the laws of physiology and pathology, together with observations in medical practice, the primary causes of disease became better understood. It was found that many diseases originated in filth, bad air, impure water, foul gases, poison from decaying animal and vegetable matter, etc. It was found, moreover, that the spread and fatality of certain diseases could be very much controlled by isolation, by disinfectants, and by a resort to sanitary laws. In carrying on this movement it became necessary to employ agents, to enlist legislation in enacting laws, and in establishing boards of health. In this way a powerful agency — sanitary science — has been brought into exercise, and under new legislation State Medicine has been introduced, the leading object of which is to prevent disease.

The more the laws of health and life are studied, the greater interest will people take in this class of subjects. In this way they will find out what are the causes of disease, and what are the laws that govern them, and that it is for *their interest* to observe and obey these laws ; also, with this study, people will learn the advantages of a sound constitution — that such an organization is more free from weakness and disease, and that in a healthy body the mental faculties seldom become deranged. Now, let insanity in all its various forms be brought to the test of sanitary science. It will be found that all its manifestations result from the violation of the principles of this science.

V. *Insanity a Disease.*— One of the most marked evidences of the progress in the knowledge of insanity is that its primary causes are traced more and more directly to the body. Says the late Sir James Coxe, than whom there can be no higher authority : —

" Insanity is a disease of ignorance,— ignorance of the human organism and the laws that regulate it; and the only way to check its growth is by a general diffusion of a knowledge of these laws, and the use of all those means necessary for the preservation of good health. Insanity originates in some form of disease, in a deterioration of the body rather than in an exclusive affection of the nervous system. The six leading factors are dissipation in various forms, over-work, meager fare, lack of ventilation, and neglect of moral culture."

In these few words we have much truth expressed. If the preservation of good health checks the growth of insanity, let the principles of sanitary science be cultivated more and more, and be brought to apply in every possible way for improving the health of the people. Just so far as it does this, it perfects the human organization and removes or moderates the primary or predisposing causes of insanity. Let these sanitary inquiries be applied to the brain, and the laws that regulate the mind. Let special inquiries be made in regard to those particular weaknesses, tendencies, or diseases, which are most likely to affect the brain. It is well understood that complaints involving the nervous system are more apt to disturb the mind. There is one thought in the paragraph quoted which should be emphasized — that ignorance of human organization is one of the most fruitful sources of insanity.

As a means to promote health of the body and sanity of the mind, it is important, then, that a knowledge of physiology should be diffused more widely.

It should be made a regular study in the family, in the school, and in all the higher institutions of learning. In the study of physical organization, we learn one fact of vital importance, that sustains a most intimate relation to the prevention of mental disease. It is this : there are differences in physical constitution, certain individuals and families being far more subject to diseases of the body and the brain than others. We find persons and families who scarcely ever suffer in body or mind, and this exemption may extend through several generations ; on the other hand, we find families whose members are subject to nearly all kinds of diseases and mental disturbances.

VI. *Hereditary Influences.*— No fact connected with insanity is more firmly established than that it largely originates directly from inherited tendencies; and, if we include weaknesses, imperfections, and diseases arising from the same source, it may be found that more than half the insanity of the present day can be traced directly or indirectly back to hereditary sources. By careful study and observation it is not difficult to discover the physical differences and hereditary tendencies in the families here described.

Let it be understood, more and more, that disease and insanity come mainly from inherited causes; let young men and women become thoroughly acquainted with such facts, and it must lead to greater carefulness in forming matrimonial alliances.

When the community is generally informed on this subject, inquiries will at once be made as to the health, the constitution, and the inherited tendencies of candidates for marriage. Such inquiries are already made in a quiet way, and they must increase in the very nature of things.

In the prevention of disease and insanity, then, heredity has a powerful influence.

VII. *Causes of Insanity.*— In the last quotation from

Sir James Coxe, is a summary of the primary causes of insanity, from one who had made the subject a special study for over twenty years.   Says Sir James, the leading factors are "dissipation in its various forms, over-work, meager fare, lack of ventilation, and neglect of moral culture."   It will be seen that each one of these covers a great deal of ground.   Passing by the last point — neglect of moral culture — the other four constitute the chief sources of disease of all kinds, some of which terminate in mental derangement.   But nearly all these great agencies, productive of so much disease of body and mind, are subject to human control, and can be more or less checked, if not entirely prevented.

The first-named, dissipation, is a fruitful source of insanity.   This may consist in drinking habits, in the use of tobacco and opiates, or in the abuse of the sexual organs by licentiousness and solitary vice.   These evils are all the results of voluntary acts, the work of a free agent ; and so they can be prevented.

Over-work of body or mind not infrequently brings on mental derangement.

Meager fare and bad air are evils which multitudes of poor people can not always escape.   Neglect of moral culture is an evil directly connected with the choice of individuals and the state of public morals.   It is a sin or an evil which can be corrected, wherever the fault may be, and there certainly can be no necessity or justification for any neglect.   Dr. Henry Maudsley, the distinguished foreign alienist, speaks on this point as follows : —

" It is to the perfecting of mankind by the thorough application of a true system of education that we must look for the development of the knowledge and the power of self-restraint, which shall enable them not only to protect themselves from much insanity in one generation, but to check the propagation of it from generation to generation. Unhappily, we are not yet agreed as to what should be the true aim and character of education."

Doctor Maudsley, here speaking of "perfecting mankind," says that it can not be done till we have a "true system of education." The only way it can be done is through the body and the brain, and to do it we must also have some standard before us, some guiding principle to aid us. As to the "propagation of insanity" by hereditary influences : how can we understand the laws of inheritance unless we have some standard in physiology? When the laws of this science are fully understood, it will be found that the most powerful agencies for preventing insanity lie in this direction.

Again : It is well understood that the most favorable time to cure insanity is in its first stages ; on this account, it is constantly urged that all insane persons, just as soon as any marked symptoms of the disease appear, should at once be sent to a lunatic hospital. This counsel has generally prevailed in acute and violent cases, but in the milder forms of the disease the friends frequently object and delay. It is a great step to take ; there are certain forms of law which must be complied with; then, the dread of its effects on the patient, the trouble attending the removal, and the anxiety about the situation and treatment of the patient in the hospital, etc.,— all these things cause delay, sometimes for weeks and months, and may prevent the patient from going till the acute stages of the disease are passed. The complaint is often made by superintendents that large numbers are sent to the hospital who can not be cured because they come too late. This is given as one of the reasons why the rate of cures is so small; for, taking all admitted into our hospitals, only about forty per cent., on an average, actually recover.

VIII. *The Medical Profession and Insanity.* — It is unfortunate that more than one-half of all the insane availing themselves of the advantages of a hospital for cure, must settle down into a chronic form of this disease, with very little chance of ever being cured. Now, suppose the members of the medical profession were so well acquainted with the diseases of the nervous system and the brain that they could detect the first symptoms of disturbed or deranged states of the mind ; they could then treat them understandingly, and in many instances successfully.

As things now are, physicians in regular practice do not take much responsibility in such cases, seldom prescribe for them, and seem quite willing they should be transferred to a lunatic hospital. This course is unfortunate for all parties; certainly for the prevention of the disease. Medical students, while preparing for the profession, should be educated to understand the diseases of the brain, as well as of the heart and the lungs; should have such a general knowledge of insanity (especially of its early stages) that they could not only detect its first symptoms, but, from knowing personally the peculiar organization and diseases of the individual and the family, they should so skillfully and easily manage the patient as in many cases to prevent confirmed insanity.

In this way large numbers might be prevented from becoming permanently insane.

Let us see what superintendents themselves say on this subject. One who has been superintendent for many years of the largest and oldest state hospital makes this statement: " Lunatic hospitals do not prevent insanity, because they do not, by the intercourse of their officers with society at large, by their published reports, and by their general relations to the public, seek to enlighten the people on the subject of insanity, its predisposing causes, its hereditary tendencies, its relation to intemperance, poverty, and crime ; and therefore they do not improve the community in this respect, except in removing from its care some of its greatest burdens."

Says another expert, who was long a superintendent of one of the largest hospitals in this state : "The more we see of mental disease in its various forms, the more are we convinced that the study of its *prevention* is infinitely more important than even the study of its *cure ;* and that the dissemination of more correct views of the true way of living, and a more rigid observance of the laws of health and Nature, would greatly diminish its frequency."

Says the superintendent of another hospital in his report: "For the treatment of insane persons we could wish some practice more encouraging in its remedial effects might be devised. As now administered, asylums for such unfortunates afford little more than a place where they may be isolated from society, kindly treated, and a watchful oversight maintained to prevent them from committing injury upon themselves or their attendants."

Let the same course be pursued in reference to preventing insanity that has been employed to suppress other diseases. Ascertain the causes, and diffuse information. This may be acomplished in a variety of ways : by enlisting the press, through books and journals, by family and educational training, by legislation, and by associated action.

Vital Statistics.

————•————

THE term vital statistics is very comprehensive in its meaning. In a limited sense it may apply to the history and life of an individual, and in a larger sense, to the subject of population, with all its changes. There is a more common use in its application to those events in life, under the heads of births, marriages, and deaths. The first act in Great Britain creating such a department, passed Parliament in 1838, and its reports have been continued annually since. The first department of the kind established in this country, was made by the Massachusetts Legislature in 1842.

The registration of births, marriages, and deaths affords valuable materials not only for personal history, but for improvement in society. So important are the mere date and place of birth and death regarded, that they constitute almost the only memorials placed upon the casket at burial, or upon the tablet which marks. the spot where the body finally rests. The preservation of precise facts connected with birth, marriage, and death — continued for many years — furnish the data for establishing most important principles bearing upon health and life. Thus the average age of man in different countries is found, whence are deduced the tables for life insurance. Thus, too, the salubrity of different localities or regions of country, the prevalence of particular diseases in different places, the effect of different pursuits or occupations upon health and life, are ascertained.

The collection of the leading items respecting birth, marriage, and death — especially of the latter — constitute the first steps in sanitary improvement and legislation. Such materials furnish the data for analyses and comparisons of one nation or people with another, in respect to the laws of health and life. The more exact and complete and the longer continued these collections are, the greater their value and usefulness.

Such registration or collection of facts is indispensable not only as a basis to initiate sanitary reforms, but to enable Boards of Health to prosecute their work successfully. Hence it should be a primary object in every community or state to gather up these statistics carefully, and preserve them in some convenient and permanent form, where a good use can be made of them.

An important element in *vital statistics* is the question of increase or decrease of population, and what are the causes. The matter of birth and death enters largely into the inquiry, and thus the whole subject of registration, in all its bearings, must be taken into the account. It is found that we can not make a proper application of the facts gathered by registration, except as based upon population, and then the changes, both in number and character, of this population must be considered. For illustration : the birth-rate in one race or class of people may be much larger than in another; so there may be also a marked difference in the death-rate. Then there may be constant changes going on among a people by emigration or immigration, which must be carefully scanned before reaching reliable conclusions. Thus several distinct points are involved in the discussion.

Formerly these reports in Massachusetts divided the population into two divisions, foreign and American. Changes by increase or decrease could be easily traced. It was understood, then, to be by descent or nationality.

But for quite a number of years the returns have been made by nativity or place of birth.　Thus the registration report of Massachusetts for 1886 is as follows: Whole number of births, 50,788; both parents native-born, 19,531; both foreign-born, 20,758; native-born father and foreign-born mother, 4,518; foreign-born father and native-born mother, 4,781; and not stated, 1,200.　Thus exact lines can not be drawn, as the classes are mixed.

The two classes registered, native and foreign-born, are composed almost wholly of foreign descent and make 9,359 births.　These, added to the 20,758, make 30,117 of this class.　But a large number of those reported "native," 19,531 belong to the same class, as their parents were of foreign descent.　No exact estimate of this number can be made, though it must be quite large, and is constantly increasing.　In Rhode Island the reports are made differently under the head of *parentage*.

A census was taken in Massachusetts, 1875, in which special pains were taken to obtain a variety of items respecting population, such as the relative number of births of American and foreign ; the size of families ; the nativities and ages of mothers ; the number and age of children belonging to each nationality ; the conjugal condition of women, etc.　Says the compiler of the census:

" The object was to ascertain the relative fecundity of women of different nationalities, and to settle, as far as Massachusetts is concerned, the question which continually arises concerning the growth of our native population as compared with that of our foreign-born.　The tables are full of instruction.

"The total number of women in this state who are or have been married is 398,759, and the number who at this date have become mothers is 309,520.　Of this number, 190,311, or 61 +per cent., are native-born mothers, and 119,209, or 38 +per cent., are foreign-born mothers ;

that is, while the proportion stands 74 + native and
25 + per cent. foreign, the mothers are 61 + per cent.
native and 38 + per cent. foreign. Of the whole number,
— 631,131 — of native-born females, 190,311, or 30 + per
cent., have become mothers ; while of 222,825, the whole
number of foreign-born females, 119,209, or 53 + per cent.,
have become mothers. Of the 89,239 married women in
the state having no children, if the line could be drawn
showing what proportion belonged to each class, there are
good reasons to believe that by far the large majority
would be found in the purely American. The average
number of births to native-born is given as 3.52 ; to foreign
mothers, 4.91 ; Irish, 5.03 ; Canadian, 4.78 ; English, 4.40,
and German, 4.23."

But all these figures from the Massachusetts census are
calculated to mislead us in our comparison, inasmuch as
they are based upon birth-place, and not on parentage. If
all that are actually foreign in *descent*, here classed as
American simply because born in the United States, could
be eliminated, it would change materially these figures,
diminishing the per cent., under the head American, and
increasing that of the foreign.

Let us take another point of view — viz., the *size of the
family*. The census shows the number of births to each
married woman, commencing with the native-born mothers,
a large number of whom have only one, two, and three,
while the foreign-born hold out, in a far greater proportion, having four, five, six, seven, and extending to ten or
twelve and occasionally upwards. The tables reporting
these facts in two columns, side by side, exhibit quite a
striking contrast.

From the first settlement of New England the birth-
rate was high, and large additions were made every year
to its numbers. This continued for several generations ;
parish records and genealogical histories showed large

families up to the commencement of the present century. But the birth-rate and the size of the family have been steadily falling off for two generations. In fact, the birth- and death-rates have been approximating nearer and nearer to each other in the strictly New England stock, so that the actual increase in numbers from this source, in many localities, is comparatively small. If the theories of Malthus on population, instead of being based upon the birth-rate in the earlier history of New England, which doubled the population once in twenty-five years, were applied to the present or last generation, it would make a surprising difference; in fact, it would find no support.

In this steadily declining birth-rate for nearly a hundred years among our New England people, in the present marked differences of fecundity between the American and foreign, and the striking changes taking place in our population, some important questions are involved. These changes do not come by accident or chance, but must have causes — causes adequate and substantial, though radical and becoming somewhat permanent. Instead of being satisfied with this abnormal state of things, let us address ourselves earnestly to finding out these causes, inquiring what laws in vital statistics have been observed, and what violated.

The causes usually assigned for decline in birth-rate and decrease in population, such as want of food and employment, bad climate, disturbances in government, wars, epidemics, earthquakes, etc., have not here occurred, and therefore can not be applied. The causes, upon a superficial survey, are not apparent, are not easily found — but still there must be *causes*. We venture to express the opinion, that those causes arise mainly from certain changes and differences in physical organization; that there is a normal standard of physiology, which establishes

13

a general law of propagation, and that deviations from this
standard affect not only the conditions, but what are considered the laws of life and health; and, if these deviations are marked in certain directions, they affect the birth-
rate and, of course, the increase of population.

Thus, in comparing the fertility of the different races,
living side by side, and surrounded by the same conditions
of climate, food, government, etc., what makes the difference in birth-rate? Why should the birth-rate of New
Englanders fall off so much in the course of a hundred
years? Why should it now be only one-half that of the
Irish, the English, the German, the Canadian-French?
The facts presented in the census and registration
reports, demonstrate beyond controversy that there is
near this difference. It should be borne in mind that this
comparison is based not upon population as a whole, where
the conditions may be unequal — that is, the American
having a disproportionate number of aged and single persons, while the foreign is composed largely of persons from
fifteen to forty-five years of age — but the comparison is
made upon strictly the married women of both classes; not
for one or two years, but for a series of years; not upon
one or two families, but upon a large number of families.
The comparison is based upon people living not in cities
or towns alone, but upon all classes of married women living scattered in both cities and towns. It would seem as
though no fairer or more correct analyses and comparisons
could be instituted.

In the investigation of such questions as those which
have here been raised, something more is involved than
the gratification of an idle curiosity, or the support of a
favorite theory. Thoughtful minds are inquiring what is
the cause of these changes, and what do they portend?
What laws in vital statistics have been observed, and what
violated? What lessons do they teach in respect to in-

crease or decrease of population? May there not be some problems to be solved, some new principles in science to be evolved? Time and further researches will determine. For wherever or whenever a civilized and enlightened people fail to obtain a proper increase in numbers, from generation to generation, it shows something wrong and unhealthy in their domestic relations, something defective in the type of their civilization and Christianity, some violation of the important laws of health, life, and human increase.

The Law of Human Increase.*

IT is almost one hundred years since the attention of
T. R. Malthus was first called to the subject of population and its changes. As his views have had more influence than those of any other writer, it is well to notice
briefly what they were. His leading principle is, that
"population, when unchecked, increases in a geometrical
ratio, while subsistence increases in an arithmetical ratio."
He held that "population is necessarily limited by the
means of subsistence," and "invariably increases where
those means increase, unless prevented by some very powerful and obvious check." He divides these checks into
two classes, the positive and the preventive. Among the
former are wars, famine, diseases of all kinds, unhealthy
occupations, extreme poverty, great cities, etc.; and in the
latter class are abstinence from marriage and sexual intercourse, from considerations of prudence. The last class
come more directly under the control of human agency.

The next writer of any note was Thomas Doubleday,
who published in 1840 a work with this title: "The true
law of population shown to be connected with the food of
the people." The term "true law" was undoubtedly introduced in opposition to the doctrine of Malthus.

Doubleday attempted to demonstrate that "wherever
a species or genus is endangered, a corresponding effort
is invariably made by Nature for this preservation and continuance, by an increase of fecundity or fertility; and

* From the *Popular Science Monthly*, November, 1882.

that this especially takes place whenever such danger
arises from a diminution of proper nourishment," and that
consequently "the deplethoric state is favorable to fertility."   Thus, "there is in all societies a constant increase going on among that portion of it which is the
class worst supplied with food—in short, among the
poorest."

The April number of the *Westminster Review* for
1852 contained an elaborate essay by Herbert Spencer,
introducing a "New Theory of Population," deduced from
the general law of animal fertility.   He "maintained that
an antagonism exists between individualism and reproduction ; that matter in its lowest forms—for instance, of
vegetables—possesses a stronger power of increase than
in all higher forms; that the capacity of reproduction in
animals is in an inverse ratio to their individuation; that
the ability to obtain individual life and that of multiplication vary in the same manner also, and that this ability
is measured by the development of nervous system."

Fourier and some French writers have advanced the
idea that "just in proportion as individuals become advanced in civilization, in the same proportion the race in-
clines to run out "; but whether this depends upon some
change in physiological laws, or upon the influence of external agents, we are not informed.   In establishing any
law or general principle, it is highly important to understand distinctly what this principle is and its basis.
During the present century, the above-named persons are
almost the only writers who have proposed any thing like
a general law or principle to guide the growth and changes
of population.

The principle laid down by Herbert Spencer is the only
one based strictly upon physiology.   All the discussions
and views of Malthus and Doubleday depend mainly upon
food, climate, government, state of society, epidemics, war,

etc.   They make the leading factors, the primary agents in all these changes, outside and in a great measure independent of the body.   It would seem more consistent with common sense and all natural phenomena, that the law which governs the existence, growth, and changes of a living being should have its basis and development in that same organization.   The truth of this principle is strikingly illustrated in the changes that have taken place in domestic animals.   The human system can not be made an exception to a universal principle.

This law of increase or propagation — the most important of all laws — must, in the very nature of things, be inherent in the body; must be incorporated into its very existence, though in its operations it may be affected by extraneous causes and influences.   However powerful may be the effect of climate, food, and other external agents upon the application or working of this law, whether to impede, thwart, or modify its operation, the law must exist, we believe, in the body itself, and in a great measure control it.   The various changes to which the human body is subjected, can not happen by chance or accident; neither can the causes be dissimilar or contradictory in different nations and races; neither can they radically change or vary, from one generation to another. Universality and unchangeableness must characterize such a law.   The reason why correct principles have not been brought to bear more directly upon the growth and changes of population is, that the principles of physiology were not formerly understood.   The science was scarcely known at the time when Malthus and Doubleday published their works, that is, the principles of the science in many of its most practical applications.   In fact, it may be safely said that some of these principles, as far as their application is concerned, are still in their infancy. One of the most interesting and important of these applications will be found, we believe, in establishing a general law of human increase.

After years of observation and reflection, we have been led to believe that there is such a law, based on physiology, and propose to submit some of the facts and arguments upon which this belief is based. As the subject is so vast and complicated, a large volume would be required to discuss it properly; we can present here only a few points or topics, by way of argument and illustration. In order to present a clear and connected view in a short paper, few quotations or references will be given.

What, then, is the briefest definition that can be given of this law? *It consists in the perfectionism of structure and harmony of function;* or, in other words, that every organ in the body should be perfect in its structure, and that each should perform its legitimate function in harmony with all others.

The nearer this ideal standard is reached, the more completely the law of propagation will be carried out. Such a basis harmonizes with the fundamental or general laws of Nature, as we find that they are based upon the highest or most perfect development of her works. Any other basis or lower standard would reflect upon the Creator of all things, and interfere with the harmony and order which exist in Nature's operations. Thus, in reference to every organ in the human body, there is such a thing as a normal, perfect structure, and, wherever this exists, they constitute a perfect model or standard of the whole system. All diseases interfere at once with the operations of this law, especially those that are considered hereditary. This class of diseases change with each generation, and sometimes become so intensified that they impair the vitality and strength of the system to such an extent as to prevent propagation. There is a class of diseases or weakness, described under the head of "sterility," "barrenness," and "impotence," from which strong evidence may be deduced in proof of a general law of increase.

There is a law in physiology, favorable to this theory, described by Doctor Carpenter thus: "There is a certain antagonism between the nutritive and reproductive functions, the one being exercised at the expense of the other. The reproductive apparatus derives the materials of its operations through the nutritive system and its functions. If, therefore, it is in a state of excessive activity, it will necessarily draw off from the individual fabric some portion of aliment destined for its maintenance. It may be universally observed that, when the nutritive functions are particularly active in supporting the individual, the reproductive system is undeveloped, and *vice versa*."

Let, therefore, on this principle, any class of organs or any parts of the body be unduly or very much exercised, they require the more nutrition to support them, thereby withdrawing what should go to the other organs. In accordance with this physiological law, if any class of organs become predominant in their development, they conflict with this great law of increase. In other words, if the organization is carried by successive generations to an extreme, that is, to a high nervous temperament—a predominance of the brain and nervous system — or, on the other hand, to a lymphatic temperament — a predominance of the mere animal nature — it operates unfavorably upon the increase of progeny. Accordingly, in the highest states of refinement, culture, and civilization of a people, the tendency has always been to run out in offspring; while, on the other hand, all tribes and races sunk in the lowest stages of barbarism, and controlled principally by their animal nature, do not abound in offspring, and in the course of time they tend also to run out. The truth of both these statements is confirmed by history.

The same general fact has been observed among all the abnormal classes, such as idiots, cretins, the insane, the blind, the deaf-and-dumb, and to some extent with extreme or abnormal organizations, such as are excessively corpulent or spare, as well as of unnatural size, either very large or extremely small.

It would seem that Nature herself determines to put an end to organizations that are monstrous, that are defective and abnormal or unnatural or imperfect in any respect. All history, we believe, proves that such organizations are not prolific, and then umber of this class born into the world, reaching an advanced age, is comparatively not large. Such facts would indicate that there must be a general law of propagation that aims at a higher or more perfect standard.

If this principle is applied to distinct classes in society, some striking illustrations may be obtained. Take the families belonging to the nobility, the aristocracy, or the most select circles, where by inheritance, refinement, and culture the nervous temperament has become very predominant: it is found that such families do not increase from generation to generation, and not unfrequently, in time they become extinct.

A similar result has also followed the intermarriage of relatives, from the fact that the same weakness or predispositions are intensified by this alliance. On the other hand, if these relatives have healthy, well-balanced organizations — even if they are cousins — they will abound with healthy offspring, and the stock may improve, and not deteriorate, from the mere fact of relationship.

Again: If we take those families and races which for several generations have steadily increased most, we shall find that, as a whole, they possess a remarkably healthy, well-balanced organization. Illustrations of this type we shall find abound most among the middle or working classes of the German, the English, the Scotch, the Irish, and the Americans. The strictly native New Englanders are, in some respects, an exception and require a more particular notice.

During the last century the colonists of New England, made up mostly of English stock, multiplied rapidly. So great was their natural increase that they doubled in numbers in less than twenty-five years. Malthus regarded them as the best specimens, in this respect, of any people or race, and based upon facts from this source his great principle of population. But a most surprising change has taken place within one hundred years, with this same people. From records carefully kept, it appears that the average number of children to each family has decreased with every generation ; that they commenced with large families — averaging eight or nine — but it is now doubtful whether the average will exceed three children a family, scarcely enough to keep the original stock good in numbers. This change has occurred in the same places, with the same people, having the same climate and plenty of food. Making allowance for the "arts of destruction and prevention" which may exist to some extent, we do not see how this great decrease in birth-rate can be accounted for, except by some change in physical organization — and this fact is true of the women as well as of the men. But a great change in this respect has taken place.

The men are not so strong and vigorous as their grandfathers and ancestors, and the women have deteriorated physically in a surprising degree. A majority of them have a predominance of nerve-tissue, with weak muscles and digestive organs. The most marked change in this one hundred years, in organization, is the loss of balance or harmony in the organs, and especially in women it is far more striking. They have been diverging more and more from that normal standard upon which the law of propagation is based.

There is only one other people or race where there has been such a natural decrease in numbers — that is, the Sandwich-Islanders. Once they were a strong and robust people. In 1830, when the first census was taken — which was ten years after the American missionaries commenced their labors — the population was 130,000, but by the last census there were only about 40,000, one-third as many as fifty years ago. In the mean time religious institutions have been introduced, education has become general, and the family as an institution has been established. All the elements of a Christian civilization have been thoroughly established, but still the population has been steadily decreasing at the rate of about one thousand each year. How can this be explained ? It can not be from the want of food, nor a well-regulated society, nor change in climate, nor want of a good government; there have been no wars, no famine, and only two or three epidemics, which were quite limited. The cause of this loss of population can not arise from any external condition or agents, but from some law growing out of and governing the physical system. It is well known that certain diseases, resulting from licentiousness and intemperance, have been brought by foreigners to these islands, causing a physical degeneracy in the people. So powerful and far-reaching are the effects of these diseases that neither the family, nor education, nor Christianity, can eradicate them. The law of propagation has been violated to such an extent that it threatens the extinction of that people.

The laws of hereditary descent afford strong evidence in favor of some general law of propagation. The fact that "like begets like," subject to certain variations and conditions, can not be called in question. The union of two agents, possessing similar and dissimilar qualities, constitutes an important condition to which this law of propagation is subject. While it may be difficult to point out, in all cases, the exact results of hereditary influences, still it has been demonstrated on a large scale that, in the aggregate, there was the most unquestionable evidence of such agency, and that it was minute and extensive, and continued for successive generations. Now the same evidence that proves the existence of hereditary agency, implies that there is somewhere a general law, of which each and every part of this agency is part and parcel ; and no one thing will throw so much light upon this whole subject of inheritance as the recognition of a general law of propagation, based upon a perfect standard in Nature. Without such acknowledgment, all these hereditary agencies are an enigma. When this branch of physiology becomes thoroughly understood, hereditary influences will more readily be traced back to their primary sources, as well as to the secondary causes which serve at times to change and modify them. In this case, far more intelligent and efficient means will be employed to improve the race.

Again : Powerful arguments in favor of this theory of increase may be deduced from woman's organization. It is a settled fact, that the primary organism of her nature is the production of children — that by this course her average health is better, and the mean duration of life is longer. Hence there must be one type or standard of organization better adapted for this purpose than all others. We maintain that the perfect structure of her whole body and the harmony of function in every organ constitute this normal standard of increase. The truth of this assertion, we believe, can be demonstrated from four distinct points — all most intimately connected with human increase: 1, In case of pregnancy a woman with this organization suffers the least. It is well known that this change frequently brings on many complaints, and sometimes serious diseases. The more the body or certain organs deviate from the normal standard, the greater the disturbance and suffering. 2, At the time of confinement, or in the process of delivery, a woman with this organization suffers less — passes through all its stages safer, and recovers from its effects quicker and better — than those having a different organization. 3, In the matter of nursing offspring, which constitutes a very important part of child-bearing, this healthy, well-balanced organization is very necessary. The fact that only about one-half of the New England women can properly nurse their offspring is very significant of some change of organization — that there is a failure in the development of the mammary glands and the requisite power of the digestive organs—and this incapacity for nursing is constantly increasing. And, in the fourth place, the difference in the physical character of offspring is very significant. This is determined in a great measure by that of the mother. The more healthy and perfect her organization, and the better the balance of all her organs, the sounder and the more perfect will be the development of her offspring. The health and life of the child demand it.

This theory of human increase derives strong evidence from an analogous law in the animal and vegetable kingdoms. It is well known that great improvements have been made within the present century in domestic animals, and this, too, by the application of physiological laws. To such an extent have the results of observation and experiment been here carried, that this process of change and improvement has been reduced almost to a science. The terms here used — "pure blood," "thorough-bred," "pedigree," "breeding-in-and-in," and "cross-breeding" — may all be explained by two great leading principles. One is a general law of propagation, based upon a perfect standard ; and the other is the law of inheritance, subject to certain conditions. The three first-named terms have originated more from an observance or carrying out of the first law — breeding from the best stock; but the two latter terms depend more upon the effects of inheritance. The results of the experiments in improving domestic stock indicate clearly that there must be some settled rules or laws in the process; and, if so, is there not some general law governing and controlling all others? A similar law of propagation exists in vegetable physiology. It is a fact well attested by gardeners that, in order to produce flowers and fruit, the soil must not be too rich nor too poor; if the plant or tree grows too luxuriantly, its branches or roots must be pruned; while, on the other hand, if unthrifty, it must receive better culture, and its roots be enriched, before it will become fruitful. It is well understood by gardeners that, in order to raise the best fruit and vegetables, the fairest and best-looking seed must be selected. So in setting out plants and trees the best-looking and well-balanced specimens are always selected. Other facts and illustrations might be cited from this source, to prove that some general law governs in the growth and changes of organic life.

Again: Arguments in favor of a general law of increase may be deduced from three other important points in physiology. Where do we find the highest measure of or the most perfect health? It is in this same normal standard of physiology and the nearest approaches to it. In some respects the human body resembles a complicated machine: the more perfect the structure, and the more nicely adjusted are all the parts of the machinery, the less likely is any one part to get out of order. And when one part, however small it may be, gives out or breaks, it at once involves the other parts, all of which must more or less suffer. Thus the individual, the family, the people, who possess by nature the soundest and best-balanced organizations, will have, other things being equal, the greatest aggregate amount of health. Not only this, but they will secure the longest lives. This same standard of physiology, then, affords the material upon which the law of longevity is based. A careful examination of the organization of all those persons who reach a great age, we believe, will demonstrate that they naturally possessed a remarkably and evenly balanced constitution.

Again : Whenever physical standards of human excellence or models of the best specimens of the race have been sought or adduced, they have exhibited this harmonious development. The Apollo Belvedere and the Venus de' Medici represent a beautiful, symmetrical organization ; and, the nearer all parts of the body approximate to this standard, the greater is the attraction and the more beautiful the form. If there is a form or type of organization in the human species more beautiful than any other, is not this mode the standard? We believe the Creator of all things has established in physiology such a standard of taste and beauty, and that this same normal standard, upon which the law of increase is based, comprises that beautiful form or standard of taste for the human body which, it has been admitted, exists, but is nowhere well defined.

Again: Arguments in favor of this theory of increase may be deduced from the writings of Charles Darwin. Two of his leading doctrines are " natural selection " and the " law of variability." The former doctrine may be defined thus : There is an inherent principle in Nature, amid all its laws and changes, for betterment, for improvement. The same result has been found out from long experience, — that the character of domestic animals can be improved by selecting the most desirable qualities and by avoiding all that conflict with these. This principle is most strikingly manifested in all organic beings, in their constant "struggle for existence," and is happily expressed in the phrase often used by some writers : the "survival of the fittest." We believe this same principle not only harmonizes with, but is nothing more nor less than, a great general law of increase, based upon the perfectionism of all organization and harmony of function; and what are denominated "laws of variation " may be explained by the laws of hereditary descent. When we take into consideration the fact that the true law of propagation is based upon a perfect standard in Nature, all changes or deviations from that standard or model result from what are properly called laws of inheritance. With this explanation it will be seen at once that a wide and varied field is laid open for their operations, depending not only on the body itself, but upon external agencies and conditions. But the question arises, Why this "natural selection," why this "struggle for existence," and why the " survival of the fittest "? Do they not arise from a universal law in Nature, which gives to those possessing this organization in the highest degree the advantage over others?

What is this inherent principle in Nature, ever aspiring for betterment or improvement? What are the secret forces everywhere predisposing in this direction? Is there some general, universal law incorporated into organic life which favors such qualities? As this law is primarily based upon a higher or more perfect standard, all its inherent or predisposing forces have an upward or improving tendency. Thus, all who are so fortunate as to possess an organization of higher grade or better than others have certain advantages. In this way the doctrine of natural selection may be readily understood, and the survival of the fittest.

This general law, applicable to all organic beings, resembles in some respects that principle found in the human system called *vis medicatrix*. It was early discovered by physicians that, in case any part or organ in the body became injured or diseased there was a surprising recuperative power in Nature of healing or curing. All the sound parts of the body seem to conspire together to help the part or organ affected. This influence to assist seems spontaneous and always healthful. So it is with this law of propagation — it is not only conservative, but improving to all possessing more than an average share of the inherent forces of this law.

Connected with this law of population there are several points worthy of careful consideration. While it possesses a sure and permanent foundation, there are a flexibility, an elasticity, which are self-regulating, and display a divine wisdom and power. Such is the nature of this law that, in all its varied operations, it does not interfere with the choice and free agency of man. When the character of this law is fully understood, what on the one hand are the penalties attached to the violation of any part of it, and, on the other hand, what are the rewards for its observance, it presents to man the strongest possible motives for his own improvement and the advancement of human happiness generally. If man is created a free moral agent, accountable for all his acts, the law providing for the propagating of the species should certainly be of such a character that he can clearly understand its nature and sanctions. According to those theories on population where its increase and changes depend mainly upon external agents, man is made, in a great measure, a mere passive agent, having but little control or responsibility in all those important matters.

If the theory here advanced is the true law of human increase, it is not a mere theory or an abstract general principle, but is capable of almost endless applications, far more than can be enumerated. It will enable us to understand far better the nature of man, his duties and responsibilities in relation to himself, to the family, to society at
14

large, and particularly to his Maker. It will furnish us a guide or a great principle by which certain practices and fashions in society, certain modes of education, systems of morals, acts of legislation, etc., can be tested. It will throw new light upon what constitutes the true grounds of human progress and the real sources of an advancing civilization.

Heredity: A Source of Pauperism.

IN the second report of this board,* it was stated that one of the sources of pauperism is an "inherited organic imperfection of the body, a vitiated constitution, or, in other words, poor stock." Since that statement was made, our conviction of its truth has been confirmed more and more by careful inspection and inquiry for fourteen years among the inmates of the State institutions. While many of these persons, by indolent and vicious habits, have contributed to their own degradation, still there were predisposing causes in their constitutions which had a powerful influence in the same direction. Many inherited feeble bodies and weak minds, with vicious propensities. They commenced life amid circumstances most unfavorable for developing the better qualities of their nature. In attempting, therefore, to ascertain every thing that has formerly contributed to make these persons paupers, criminals, or lunatics, we must not only take into account their own agency, but must consider well the nature and amount of capital which they had to start with in life.

No fact in science is better established than that there is a most intimate mental as well as physical relation between the parent and the child, — between each generation and the succeeding one. This relation has been well expressed in the proverbs, "What is bred in the bone can not be whipped out of the flesh," and "Like begets like."

* From the Fourteenth Annual Report of the Board of State Charities, which was prepared by the writer as chairman.

Heredity has, we believe, a far greater agency in producing social evils than has generally been supposed.
This influence extends, by transmission, not only to the
form of the body and the features of the countenance,
but to every part of the system, — to the quality of the
blood, — especially to those vital organs which give
stamina of constitution and beget mental predispositions.   Whatever agencies, therefore, are calculated to
injure the body or deprave the mind, to incapacitate an
individual for self-support, or to make him a corrupter
of others, should certainly be exposed by the guardians of public charity.   Among the most mischievous
agents operating injuriously upon the human system is
alcohol ; and whether we consider the extent of its abuse,
in various forms, or the terrible effects which it produces,
it stands foremost as a cause of pauperism and other
evils.   It poisons the blood, and produces a diseased or
morbid condition of almost every organ of the body.   It
affects the brain, impairs the intellect, perverts the moral
sentiments and the will, and increases unduly the activity
and strength of the worst propensities.   It prostitutes the
higher to the lower nature of man, changing what should
be the true aims and objects of life to those of low animal nature.

Closely connected with the alcohol poison is another,
which, though not so manifest in its effects, has a most
destructive influence upon human welfare.   This poison
arises from habits of licentiousness, and its evil effects do
not cease with the living, but extend through successive
generations.

The syphilitic poison operates on the human system in
so covert a manner, and in such a variety of ways, that it
is sometimes found difficult to trace out all the effects ;
but the more thoroughly the pathological and morbid conditions of the body are brought to light by modern science,

the more extensively are discovered the mischiefs which this poison has wrought. If the amount of vice, disease, and pauperism produced from this source alone could be made known, it would surprise people. In some respects this poison is more destructive of health and life than the poison of alcohol.

But there are other modes of abusing the reproductive organs which injure most seriously both body and mind. A careful inspection of the inmates of our almshouses and hospitals will show a vast amount of suffering from this abuse. It is the hereditary effects of these evils that make them especially significant in the production of pauperism and insanity. To aggravate the matter, their effects are communicated in an intensified form, from generation to generation, and it is very difficult to check or eradicate them either by human means or through the recuperative powers of Nature.

Besides these two poisons, there are other agencies that injure the body and enfeeble the mind, such as narcotics, stimulants, over-medication, etc. Then come irregular habits, want of proper nutrition, and a train of diseases which either destroy their victims or make them helpless and dependent. When the physical system is impaired or broken down, the mental faculties frequently become enfeebled and depraved, so that not only poverty and temporary dependence, but habitual indolence and shiftlessness also, ensue. Such a state of things makes paupers, who, by natural association, form social and domestic relations with each other. The more such persons become associated together, either in families or communities, the more unfavorable is the influence of one upon another, the whole tendency of things being to sink them lower and lower in the social scale. Worst of all, whatever offspring these persons have are sure to be impregnated with vice, pauperism, and crime, by the law of inheritance as well as by the habit of association. A careful inquiry into the origin, history, and character of the inmates of our public institutions will abundantly prove and illustrate these statements.

If idleness, improvidence, intemperance, and licentiousness are prime factors in the production of pauperism, these have their germs or springs in physical organization. The desire, the craving, the predisposition, for such vices, were transmitted from parent to child. A poor physical development throughout, or a predominance of the animal nature, characterized undoubtedly the ancestry for two or three generations. A feeble, sickly body may have been inherited impregnated with disease, it may be scrofula or some other poison. The diseases thus generated are of the worst type, the most difficult to cure, and the most destructive to industry and self-support.

If this prevention of disease, or improvement in hereditary agencies, could be extended to the defective classes, — to the idiotic and feeble-minded, to the deaf-and-dumb, to the blind, and the insane, — it would make a notable difference in the amount of pauperism. The cause of our finding so large a number of paupers in these defective classes is violated physical laws, either on the part of the individuals themselves or their ancestors. While we can not determine just what proportion of these evils are of hereditary origin, nor point out exactly the line of causation in their production, there is no question but that a great deal can be done, by proper means, towards preventing them.

A striking illustration of hereditary influences was brought to light recently in the state of New York, by an official investigation on this subject. Among the results of this investigation, the history of a remarkable family is given, under the name "The Jukes," extending back six generations, where, from one bad woman, nearly a thousand persons, by birth, relationship, and association, became paupers or criminals. This history shows the great power and influence of heredity and early associa- tion, in the production of pauperism and crime, more forcibly than ever before.

It demonstrates that the seeds, or primary causes, of these evils are connected with the great laws that govern human life farther back than has been generally supposed. This family history brings up also the relations between "heredity and environment," and suggests means or agencies which may be employed to prevent or check the miseries originating from these two sources. The more this whole subject is investigated, the more evident it becomes that, in order to check the increase of pau- perism, crime, and insanity, the remedy must be applied to their primary sources. It will be found, too, that these are, to a great extent, under the control of human agency.

A careful examination of all the facts gathered on this subject shows that, in addition to the hereditary influence, ignorance, idleness, intemperance, and prostitution are prime factors that enter into the complex product of pauperism and crime. The hereditary agency precedes these personal factors, and predisposes to their activity and control. It is evident that the germs or predisposi- tions originate in physical organization and development. These secondary agencies would not be called out, or would have but little influence, if the right kind of material for their operation had not been provided.

Improvement in Domestic Stock.

THIS subject is introduced to show that the same general principles apply here that pertain to the human species. It is true improvements have been made without recognizing these principles, but by a long course of experiment and observation the same results have been reached. It is over one hundred years since these experiments started in Great Britain, but they have improved so much as to be reduced almost to a science. In this country there has not been the same interest or pains taken, neither has there been the same success.

While the general law of propagation, based upon anatomy and physiology — or perfectionism of structure and harmony of function — holds the same in domestic animals, there exists a wide difference in the operation of the laws of inheritance. Some of these laws apply, the same as in the human species, but with others there is, at the same time, a very wide difference. There are three very important points of distinction. These are radical, fundamental, and fixed. In applying this law of propagation, the existence and the influence of these points of difference must be taken into account.

The first distinction is reason and intelligence, or the intellectual and moral nature of man, — that all the faculties which distinguish him from the animal have a most powerful influence upon this law of inheritance.

The second difference is the marriage institution. Without resorting to Revelation at all for a divine sanction of this institution, we believe its necessity can be proved upon physiological laws alone, — that the health, happiness, and highest welfare of the race require just such an institution; in fact, that the human species as a whole can not be perpetuated in its highest type without the marriage relation, and that the law of inheritance must act in harmony with its sanctions.

The third distinction is in the objects of creation. Man is a free moral agent, accountable directly to his Creator for all his powers and his acts; but the animal was created with a very different nature, and for different purposes. The laws that govern his organization can be more easily applied and directed, especially by human agency. They are not only more simple and less complicated, but can be brought to bear more directly and with more immediate results. This law of heritage is here, in a great measure, not only under the control of human agency, but what may be called the physiological influence is small compared with what it is in the human species; and then the external agents, such as food, climate, and exercise, can here be directed and applied far more aptly and successfully.

It seems to be a wise provision that this law, in case of domestic animals, can be controlled very much by human agency. In this way, and by this means only, can great improvements be made. By a correct knowledge of this law, and the hereditary influences growing out of it, these improvements can be carried on far more successfully and intelligently. Experiment and observation have done a great work here, but knowledge clear and definite, based upon law, will furnish a far better guide and stimulus. Guided by these principles, the science of breeding will be clothed with new interest, will be pursued in a more intelligent manner, and with a greater certainty of securing the desired objects. Under such auspices, may we not expect that the improvement of domestic animals will be more rapid, sure, and permanent than it has ever hitherto been?

Physical Degeneracy.*

THE term degeneracy implies a decline in qualities which were once possessed, and which pertain to a higher state of being or a more normal standard. This decline may take place slowly or rapidly, and be transient or permanent in its character. The nature of the changes and results depends of course upon the subject involved. It is proposed at the present time to point out certain changes taking place in the physical systems of large numbers of our population that show a well-marked degeneracy.

There are several points from which the lines of this declination may take their origin or start. It may be from a perfect standard or organization with which man was first created, representing the soundest and highest development of all the organs in the human body, in a well-balanced state. It may also refer to the physical systems of those races and nations most advanced in civilization, or to the constitutions of the first settlers in this country, together with our immediate ancestors.

It is not our purpose to enter at all upon the domain of anthropology or ethnology, but mainly confine our observations to certain changes taking place particularly in and among people who have their origin and nativity in New England. Neither will it be possible to notice all the causes and agencies operating here to change physical

* From the *Psychological Journal of Medicine* (Appleton's), N. Y., October, 1869.

organization, much less their results; the field is too large, the subject is too complicated, and the effects are too far-reaching. And as to the numerous and varied changes in morals and in mental developments, reference can be made to them only as affected by the body. The more the laws of the physical system are examined and studied, the greater will be seen the importance attached to them, in their influence upon mental improvement and moral development.

The causes operating to produce these changes of organization are very numerous — some external to the body, and some internal. So multiplied and complex are they, that it is very difficult to describe them in detail or draw distinct lines between them. Climate has always been regarded as one of the most powerful agents in changing the physical system — especially is this the case when applied to different individuals or races in removing from one residence to another; but, where the same people continue to live in the same locality for several generations, the change occasioned by climate can not be sufficient to make any appreciable difference. Thus in New England, though there prevails an impression that some change has occurred here in the seasons, if not in the climate, since the settlement of the country, it is so slight that not much account can be made of this alone. It is true, however, that if the constitution has become sensibly impaired by other causes, and weakened in particular organs, the same climate may have a more marked influence upon it; but, even then, it could not be considered as a leading agent.

Among the external agencies may be mentioned the effect of changes in private and public institutions, in the style of dress and state of society, in the kinds and modes of doing business, in the changes of soil, of vegetation, of air, of dwellings, in methods of education, habits of domestic life, etc., etc. ; but then, many of these external agencies can not be considered separately from the internal, which may be summed up under three general heads, viz., exercise in all its diversified forms ; foods, including drinks, medicine, and whatever enters into the system ; and the last, though by no means the least, the effects growing out of the laws of hereditary descent. While many of the causes depend upon circumstances and surroundings, frequently beyond the choice and control of individuals, still some of the most efficient, such as exercise, food, and personal habits, come within the power of every individual, provided he has sufficient intelligence.

To describe in detail all the agencies, and the precise way in which they affect the human system, does not come within the range of our present inquiry, so much as what are the direct effects of their influence upon the constitution, what actual changes they produce in the body, and what will be the probable result. Neither will it be possible to examine minutely into all these changes, or describe just how they are brought about, but simply notice those more marked and important. With few exceptions, these changes in the system occur so gradually, so quietly, and so imperceptibly, that they are not noticed at the time, and are scarcely felt by the individual himself.

Some of these changes, it should be stated, are brought about principally by the exercise of the mental faculties. That the body, whether in a healthy or diseased state, has a direct and powerful influence upon mental manifestations, is admitted ; while, on the other hand, the exercise of the mind, including the animal propensities, the moral sentiments and intellectual faculties in all their diversified operations, has a great effect upon both the development and the functions of the physical system.

The brain itself, as the organ of the mind, is subject to many changes, and, as the center of the nervous system —which ramifies every organic tissue — exercises a powerful influence over every part of the body. And in proportion as the nervous temperament becomes more and more predominant, with a large active brain, the same proportion will the influence of mental exercise have over physical organization. As this predominance of the nervous system is relatively increasing every year — becoming a marked feature in the type of our present civilization — the exercise of the mind and the nerves is destined to have more and more influence over the body, whether in a normal or a morbid condition. By a constant, intense, and increasing activity of mind, those organs that contribute most to its wants will not only be developed more and more by the general law of exercise, but also by sympathy and suffering, to which this mental strain peculiarly exposes the nervous system, while, at the same time, all other parts of the body are by this means powerfully affected.

The question might here naturally arise, how, or by what process or law, is physical organization changed? To this we answer, that such changes are principally affected —

1. By the natural law of growth by exercise, nutrition, air, light, etc.

2. By positive violation of those laws which Nature has established for the growth and preservation of the system.

3. By simple neglect or want of the use of all those means necessary for the healthy growth and development of every organ in the body.

4. By disease in all its various forms and terrible results.

5. By the gradual but steady operations of the laws of

hereditary descent, working for good or ill, according as they are obeyed or thwarted.

This last cause is very fruitful in results, inasmuch as it includes all the others ; and, where two or three generations are taken into account, the changes that may be effected will be equally surprising in their nature and their extent. It is not our purpose to consider these causes now, or in their regular order, as they will come under review, more or less, in discussing the various changes that have taken place in the physical system ; but, before proceeding to this part of the subject, it may be well to have in mind some definite standard of physiology to which reference can occasionally be made.   In all the works of Nature or Art, it is a great advantage to have set before us a perfect standard, or the highest development of the class or kind under consideration, in order to in-institute proper comparisons, or make careful discrimination in the changes taking place.

When man first came from the hands of his Maker, we have reason to believe he was created with a perfect or-ganization.   Every organ was perfect in itself — without spot or blemish ; without excess or defect ; without weak-ness or disease.   Then, there was a perfect harmony or balance existing between all the organs throughout the system.   This perfection of organism constitutes in human physiology a standard upon which certain general principles or laws have their basis — their foundation.   It affords the only perfect standard of beauty, of health, of strength, of happiness, of longevity, and of increase ; or, in other words, it provides the materials whereby all these objects may be secured in their greatest possible measure, or very highest degree of development.

For the sake of illustration and convenience in reference, we will here divide all the organs of the system into four distinct classes, called Temperaments. The word *tem-*

perament is sometimes used to denote a mixture of qualities, including mental as well as physical; but, as here used, it is intended to apply only to different compartments of the body. The *first* division, including the brain, the spinal column, and nerves of motion and sensation, is called the *nervous* temperament; *second*, the heart, the lungs, and all the blood-vessels in the system, called the *sanguine* temperament; *third*, the organs in the abdomen, the stomach, bowels, liver, and absorbents, called *bilious* or *lymphatic* temperament; and *fourth*, muscles, bones, liga-, ments, constituting the motive apparatus of the system, called the *muscular* temperament.

But, unfortunately, this physiological standard, represented by a complete development of every organ in the body, and perfect harmony in all their functions, is nowhere to be found. No nation, or race, or tribe, or people upon the globe, can at the present day show perfect living examples, containing all the organs in a perfectly well-balanced state. They are only approximations to this standard.

The human constitution has been constantly changing, in every age and with all classes of people. The causes are to be found partly within the body, and partly in external agencies and influences, and are sometimes observed to vary materially with the same individual or generation. Slight changes in the organization do not affect much the physical or mental character of a people; but, when a certain class of organs, or, in other words, one of the temperaments, becomes very predominant, it has a most marked and, generally, unfavorable effect.

Moreover, if only a single individual in the community here and there was found with an organism thus imperfectly developed or badly predisposed to disease, its hereditary effects would not be very perceptible; but, when large numbers, or a majority in a community, are found so constituted, not more than one or two generations can possibly pass before such effects are generally observed and become well known.

If the standard of organization here described is strictly the normal state of man, such as he had when created and would now have in his best estate, all weaknesses, all diseases, and all imperfections of the body are abnormal, — are deviations from this standard.

It is particularly in this imperfect, abnormal, diseased state of the system, that the laws of hereditary descent come into more active operation and exert the most influence. The changes effected from this source are beyond calculation. And when the tendencies are in the downward direction, it would seem as though their forces were far more active and become intensified.

With these preliminary remarks, we proceed to consider more directly and definitely the changes in human organization that indicate a decline. No evidence of much weight can be deduced from any exact statistics in figures relative to man's physical development. No extensive or reliable collection of facts, touching the height, weight, strength, and other properties of the body, was ever gathered, that would throw much light upon this subject. But connected with our late war, Dr. B. A. Gould, under direction of the Sanitary Commission, caused examinations on these points to be made upon over a million of soldiers.

In the law of growth, Quêtelet and other European authorities have been inclined to consider that the maximum stature is not reached in our country till the age of thirty, and even then it varies with different classes of men. It was found by these examinations that the native soldiers of Tennessee and Kentucky were the tallest in stature ; next came those from Michigan, Wisconsin, and Iowa and Illinois — many of these being born in New England ; and the tallest soldiers from the Eastern states were from Vermont and Maine. The question is here raised, What particular agency or influences favor most the growth of the body in stature ? and, after examining and comparing the various theories upon this subject, Doctor Gould comes to this conclusion, namely: "That all the influences here considered — climate, nationality, comfort, elevation — may contribute in some measure to affect the stature is more than probable; that both ancestral and local influences are recognizable is certain. And although we can not succeed in determining what is the *chief* agent, it may not be without value that we furnish evidence of what it is not." One object in making this quotation is to suggest what this "*chief agent*" may be, viz., is it not *the exercise of certain muscles and bones, while in a state of growth, and living in an uneven, hilly country, requiring much use of the legs and spine?*

While our American soldiers took the lead in stature, the examinations showed that the representatives of other races surpassed our men in weight, strength, and certain other properties touching the dimensions of the body. As to physical stamina for enduring long marches, the hardships of camp-life, and other exposures incident to the war, it was not easy to make any satisfactory comparisons between the different classes.

In making examinations with reference to the draft at the opening of the war, it was a matter of surprise to surgeons what a large number of men in the community were found whom, by reasons of infirmities or diseases, they were obliged to exempt from the draft. If exact information could be obtained as to just what proportion of men, at the present day, are physically disqualified for military service, the result, we believe, would surprise the public.

This brings us to consider a most important change in the organization of our people, viz., *gradual loss of muscle*

15

and *increase of the nervous temperament.* Its leading
tendency is to diminish the stamina and vitality of the
constitution, as well as increase and intensify unduly the
action of the brain and nervous system. While this change,
at first thought, and from certain points of view, may
seem an improvement in the estimation of some persons,
yet in the end, when carried out to the extent which now
seems probable, we apprehend that it will prove serious in its results. Perhaps the truth of this statement
can not be demonstrated by the figures of arithmetic or by
exact statistics, yet we think such an amount of facts and
arguments, in variety, pertinency, and force, can be presented in its favor, as to establish it beyond reasonable
doubt.

In the *first place*, the increasing migration of our people
from the country to the city is decidedly unfavorable to
physical stamina and life. Within forty or fifty years
there has been a marked change in this respect, and every
year witnesses its increase more and more.

The desire of raising themselves in the world and of
improving their circumstances is constantly impelling
large numbers in the rural districts to remove to the city
or large town, where wages are higher, the advantages of
society greater, and the conditions of life more attractive.
With many the leading motive for change seems to be to
get rid of manual labor and hardships incident to country
life; and obtain a livelihood in the city, by means of
lighter employment or rather by their "wits" than by hard
work. The introduction of new mechanical and manufacturing business, together with the widely-extending domain
of trade, is continually encouraging this migration. To
such an extent has this change already occurred in population, that almost one-half of it in the older states is now
found in cities and large towns, and there is reason to believe that the proportion is steadily on the increase. Now

no one truth in vital statistics is better established than the fact that *city life* tends to reduce the physical energies of the body and shorten human life. The close confinement in-door, the breathing of vitiated air, the frequent use of unwholesome water, the increased habits of intemperance and licentiousness found in cities, have a pernicious effect upon the human constitution, by multiplying its infirmities and its maladies. It has truly been said that an exclusively city population would certainly run out if it were not continually replenished from the country.

In the *second* place, the very general giving up of *farm-work* and the more laborious employments, on the part of our New England people, is very unfavorable to muscular development. It is a well-known fact that a very few of our young men are willing to follow, practically, agricultural pursuits, and every year witnesses a less and less number disposed to learn or follow the more laborious trades, such as the mason, the carpenter, the millwright, the wheelwright, etc. It is reported that the superintendents or master-workmen of such trades can not now get apprentices at all to learn the business.

One of the ostensible objects at the present day is, How can we avoid manual labor or hard work? It is in part this that has led to the invention of labor-saving machines, to substitute the use of water-, steam- and horse-power, for human agency, and devise various means or contrivances for conveyance and travel. What a wonderful contrast between the work now performed here and that of fifty or a hundred years ago! What a vast quantity of rocks were once gathered up or dug out of the ground in New England, and what an immense amount of stone-wall was laid! What an untold amount of hard work was performed in clearing the forests, in subduing the ground, in cultivating the soil, and erecting substantial buildings! Such labor made strong muscles and sound constitutions.

It is now generally admitted that neither the men nor the women of the present day have the physical vigor and stamina that their parents and grand-parents possessed. They can not begin to do the work, endure the hardships, or bear the exposures of their ancestors. The constitution has changed. The strong and well-developed muscles, the large and stalwart frame, the stout and compact form, the abundant supply of pure arterial blood — all of which characterized the first settlers of this country — are now seldom found.

These facts are patent to almost any observer, but marked in the eye of the physiologist, and rendered still more striking to the physician when the change in the present type and character of the diseases is considered — so different from what they once were. While it may be difficult to define, in every case, just what changes in disease have occurred, it is admitted by the highest medical authorities that important changes have taken place, and that the treatment once required and found successful can not now be applied.

From the testimony of aged physicians as well as from the description given by different writers, it is evident that there were, in the case of our ancestors, relatively much more acute disease, far greater violence in its attacks, and a decided higher grade of inflammation, than exist at the present day. Then, they required a great amount of venesection, and the use of powerful drugs, neither of which with us is often required or very well borne. We have, moreover, a class of diseases, arising from scrofulous complaints, from general debility, and a predominance of the nervous system, which once were almost unknown. Formerly, cases of dyspepsia, anæmia, and neuralgia — each of which is now the source of much disease — were seldom found, but now are very common.

From a careful examination of the nomenclature of the

diseases in the reports of deaths in Massachusetts for the last forty years — the period since this registration was established — we find a marked change has relatively taken place in reference to different diseases. As some changes in their classification have been introduced during this time, it is difficult to institute exact comparisons as to the frequency of particular diseases; but one thing is certain : that there has been a great increase of those pertaining to the brain and the nervous system. In the opinion of some physicians nervous diseases have more than doubled. Cases of inflammation and congestion of the brain, of apoplexy, of paralysis, epilepsy, convulsions, etc., are far more common now than formerly. Once apoplexy and paralysis were thought to be confined almost exclusively to persons from sixty to eighty years of age, but now they frequently occur from forty to sixty ; and convulsions, with other diseases of the brain, have increased surprisingly, in the case of children especially.

In the *third* place, this loss of muscle and increase of the nervous temperament, together with a change in the type and character of diseases, applies with far greater force to woman than to man. It is in her case more marked, more radical, and at the same time more injurious in its results. Within forty or fifty years a great change has taken place in the early training of girls, as well as in the domestic habits of women. Once a large majority of the girls of our American population were taught early to understand and perform housework, which, combined with considerable out-door exercise, served to develop strong and healthy physical frames. From the age of six to sixteen, of the girls of that period, probably not more than half their time, on an average was devoted to school education or intellectual pursuits. In fact, this would apply to only the higher and wealthier classes, whereas the great majority of girls of that age had much less schooling than that.

It seems to be the order of Nature that the physical
system is best developed and strengthened when the person is young — when all the tissues of the body are in a
natural state of growth — and especially is this so in the
case of the muscles which constitute the moving power of
the whole system.   Now no kind of exercise or work whatever is so well calculated to improve the constitution and
health of females as domestic labor.   By its lightness,
repetition, and variety, it is peculiarly adapted to call into
wholesome exercise all the muscles and organs of the body,
producing an exuberance of health, vigor of frame, power
of endurance, and elasticity of spirits ; and to all these advantages are to be added the best possible domestic habits, and a sure and enduring foundation for the highest
moral and intellectual culture.

But what a change has there been within a short time in
the education of girls ! They are now very generally kept
in school from the age of six to sixteen, with only short
intermissions for rest and recreation. Very little attention is paid to physical development and health. They
grow up with muscles weak and soft, possessing but little
strength and vitality. The brain, together with the nervous system, is kept continually upon a strain, producing
often no doubt a brilliancy and precociousness of scholarship without the stamina of constitution to sustain it.
Hence, many girls for the want of exercise and by too
close application to studies, now early break down in health,
or bring on weaknesses and diseases which disable them
more or less through life.

And just in proportion as this training of the muscles
is neglected in youth, in the same proportion will it disin-
cline them afterwards to perform house labor, as well as
all other kinds of work which requires much exertion. At
the same time there has grown up in a portion of the
community a strange and pernicious sentiment or feel-

ing that there is some degradation attached to domestic labor, so that nearly all of it is now performed by foreign help. In consequence of this want of training or neglect of exercise, large numbers of our women do not possess that strength and firmness of muscle, that stamina and vitality of constitution, which are indispensable to sound and vigorous health. In fact, the natural law of growth and healthy development seems to have been reversed. According to physiology, this is the natural order: first, the cellular tissue, then the muscular, the cartilaginous, the osseous, and the nervous; and, inasmuch as the muscular is the moving power of all the other tissues, its proper exercise and development in childhood become all-important. Then the brain and nervous tissue come last in the order of growth, which should not be pushed prematurely at the expense of the others. Besides, it is allowed by physiologists that, in a normal state, about one-third of the blood should go to sustain the brain, and thus, in this way, one-third of the vitality of the system is consumed. It is well known that no kind of exercise uses up the vital energies or exhausts the system like that of brain work.

Now, while all these tissues are in a growing state they constantly require a large amount of nutrition for growth, but, if this premature exercise of the brain demands more than its legitimate share of the nutrition, the whole system must suffer. The supply is not equal to the demand. Hence, the natural growth and development of the various tissues of the body are more or less checked, causing a want of vitality — a deficiency of good arterial blood. Then commences early in life a weakness, a feebleness which pervades the whole system, a peculiar paleness indicating a state of anæmia, while, at the same time, there is almost uniformly a mental activity, a nervous excitability and restlessness, entirely unhealthy and unnatural.

In considering these changes and their significance or tendencies, two things should be borne in mind : 1, These violations of law occur at a period the most critical in life, when certain important changes are expected in the female organization, and when the healthy efforts of Nature should have the greatest possible encouragement; 2, The changes in the growth and organization of the female from the age of ten to fifteen determine in a great measure her constitution and state of health in after-life. It is true, physical changes occur from the age of fifteen to twenty, and sometimes very important ones; but the leading forces that shape, direct, or modify these changes depend principally upon the agencies and influences operating on the system in previous years. It should also be borne in mind that the changes formed at this period become generally *fixed*, if not *structural*, and can not be easily altered.

Connected with this want of muscle and vitality in woman, there are certain other conditions in her organization which indicate a decline. In consequence of neglect of physical exercise and want of vitality, there has arisen a general state of debility and anæmia, which is a fruitful source of disease. In proof and illustration of this fact, there has been called for, in the treatment of women, a most surprising increase of tonics, especially in the preparations of iron. It is thought, in the whole history of the *materia medica*, there has nowhere been so great a change as in the increased variety and amount of the ferruginous preparations. While it may be impossible to estimate the exact amount of this increase, it is the opinion of some physicians in long practice, as well as druggists in extensive business, that this increase, in forty or fifty years, must be tenfold or more relatively for the same population. Besides the prescriptions of physicians, great quantities of iron are put up by druggists, and are found largely represented in patent medicines. These preparations of iron are used mostly in the treatment of female diseases or weaknesses. Once they were prescribed, after the run of a fever or an attack of some acute disease, when the system had been reduced and tonics were only temporarily required to improve the appetite and the blood; but now, in almost all the ordinary complaints of women, iron, in some form, becomes an indispensable medicine ; in fact, in many cases they depend upon it from day to day, from week to week, the year in and year out, almost as much as upon their daily food. Its use has also become extensively necessary even in cases of children suffering from debility and anæmia, which would not have been required if they had inherited organizations full of life and vitality, or had been rightly trained in physical exercises, and their systems properly nourished and strengthened.

There is another practice which is having a deleterious effect upon female health, and contributes largely towards the decline and weakening of her organization. We refer to the fashions of the day or the style of dress, which changes the form of the body, compresses the chest and abdomen, thereby preventing the proper expansion of the lungs, by which the blood is oxygenated ; it obstructs the natural action of the heart, the stomach, and the bowels, and depresses more or less all the internal organs, especially those in the lower part of the pelvis, thus interfering seriously with the great laws of reproduction.

Again : Connected with this weak and relaxed state of the muscular tissue, and with the above-mentioned effects of fashion in dress, has sprung up a class of very grave complaints, which once were comparatively unknown in our country, and are somewhat peculiar to American women. We refer particularly to weaknesses, displacements, and diseases of organs located in the pelvis. Within twenty or thirty years there have been not only marked changes in the type and character of the diseases of females generally, but *this class*, comparatively new, has increased surprisingly. No one but a medical man, who has devoted special attention to this subject, can realize fully what are the nature and extent of this change, and what are the direful effects. These complaints have frequently been produced, have certainly been aggravated, and sometimes made incalculably worse, by the various means and expedients which the parties have resorted to, in order to interfere with or thwart the laws of population. It is not this class of complaints in themselves, or in their effects upon the general health, that renders them so important, but the *relations* which they sustain to the marriage institution, and the laws of reproduction. While we can not here with propriety go into details, it may suffice to state that such are the nature and extent of these difficulties as to interfere radically with the great objects of the marriage relation, as well as of domestic life. To the thoughtful and discriminating mind this point of view affords the strongest possible evidence of decline in physical organization.

There is another marked change going on in the female organization at the present day, which is very significant of something wrong. In the normal state, Nature made ample provision in the structure of the female for nursing her offspring. In order to furnish this nourishment pure in quality and abundant in quantity, she must possess a good development of the sanguine and lymphatic temperaments, together with vigorous and healthy digestive organs. Formerly such an organization was very generally possessed by American women, and they found but little difficulty in nursing their infants. It was only occasionally, in case of some defect in the organization, or where sickness of some kind had overtaken the mother, that it became necessary to resort to the wet-nurse or to feeding by hand. And the English, the Scotch, the Germans, the

Canadian-French, and the Irish women now living in this country generally nurse their children ; the exceptions are rare. But how is it with our American women who become mothers ? To those who have never considered this subject, and even to medical men who have never carefully looked into it, the facts when correctly and fully presented will be surprising. It has been supposed by some that all or nearly all our American women could nurse their offspring just as well as not ; that the disposition only was wanting, and that they did not care to have the trouble or confinement necessarily attending it. But this is a great mistake. This very indifference or aversion shows something wrong in the organization as well as in the disposition ; if the physical system were all right, the mind and natural instincts would generally be right also.

While there may be here and there cases of this kind, such an indisposition is not always found. It is a fact that large numbers of our women are anxious to nurse their offspring and make the attempt ; they persevere for a while —perhaps for weeks or months—and then fail. They find that their milk does not satisfy the child, or that it does not thrive, and they conclude there must be a deficiency in the quality of the nourishment. In many cases after repeated trials, and finding no improvement either in the child or mother, it is decided to give up nursing entirely ; while others — depending partially upon nursing — resort to artificial feeding. There is still another class that can not nurse at all, having neither the organs nor nourishment requisite even to make a beginning. The proportion of mothers that have an abundance of good milk, and can thus support the child well till time of weaning, without any artificial help, is not large. It is the opinion of some medical men of long experience and careful observation, that not one-half of our New England women, particularly in the cities, can at the present day properly nurse their offspring. Why should there be this change?

Why should there be such a difference between the women of our times and their mothers or grandmothers? Why should there be such a difference between our American women and those of foreign origin residing in the same locality, and surrounded by the same external influences? The explanation is simple; they have not the right kind of organization; there is a want of proper development of the lymphatic and sanguine tempera- ments — a marked deficiency in the organs of nutrition and secretion. You can not draw water without good flowing springs. The brain and nervous systém have for a long time made relatively too large demands upon the organs of digestion and assimilation, while the exercise and development of certain other tissues in the body have been neglected. That we have not misrepresented or overstated the extent of this defect existing in American women can be abundantly proved from the extensive sale of nursing-bottles. But let the reader — and, if he be a medical man, so much the better — cast his eye over the circle of his acquaintances among young mothers, and count up the number who nurse their offspring, then those who unite nursing with feeding, and then those who do not nurse at all, and he will be surprised to find how many will fall into the last two classes. If, as we maintain, this be owing to a decline in physical organization, what stronger proof can we have of the nature or extent of the evil? It is doubtful whether any such change of organ- ization can be found in the history of any other people.

But the defects of the system here described do not all arise from a merely negative source; there are positive evils which can not be remedied by any artificial means. In consequence of the neglect of physical exercise and the continuous application to study, together with various other influences, large numbers of our American women have altogethar an undue predominance of the nervous temperament. If only here and there an individual were found with such an organization, not much harm comparatively would result ; but, when a majority or nearly all have it, the evil becomes one of no small magnitude. While, in the estimation of some, it affords the most favorable conditions for the highest degree of intelligence, refinement, morality, and happiness of the individual, it does not harmonize so well with the laws of maternity and the interests of humanity.

Besides the inherent defects in such an organization, in not making the necessary provisions for gestation and lactation, the natural instincts of woman in a pure love of offspring and domestic life become changed : the care and trouble of children are a burden ; society, books, fashion, and excitement generally are far more attractive. The anterior lobe of the brain has been exercised altogether too much at the expense of the posterior. If the law of human increase is based upon a perfectly sound and well-balanced constitution, represented by a uniform, equal development of all the temperaments, then either extreme must be unfavorable to the propagation of the species. Such we believe is the true physiological law, and will be found verified in the history of every race, nation, and people on the globe. In discussing the changes taking place in the organization of our people, the *form* and *size* of the body should not pass unnoticed. In the loss of muscle and increase of nerve-tissue, together with diminished vitality, we maintain that there must be a gradual change going on in the *stature*, the *form*, and *size* of the body. The outline or framework of the system is made up principally of certain bones, and the growth of these is insured, at a certain age, by proper exercise and nutrition under favorable influences. The differences of various races and nations, as well as of individuals, in these respects, we think, can be explained in some measure by the application of this principle. The size and form of the body depend much also upon the proper development of the organs ranged under the lymphatic and sanguine temperaments. A predominance of the nervous system is seldom accompanied with a body of large size or structure, but is more generally found in persons of slender build and medium size. While changes in the stature, form, and size of people, as a whole, can not be determined at once, two or three generations can not pass without their becoming very perceptible.

In a former place we referred to the portly forms, large size, stout build, and strong constitutions of our forefathers. The descriptions and portraits of our Puritan mothers represent them for successive generations as possessing well-developed bodies, and in many instances of large size. Such is the testimony of elderly people generally, and they comment particularly on the diminutive size and slender form of the women of the present day. Testimony similar to this has often been borne by foreign tourists visiting our country, especially in respect to our women. This change in the constitution is more important as applied to them, on account of the laws of hereditary descent. While we may not be able to demonstrate in figures that such changes as are described above have actuallly taken place among our people, we do maintain that there are causes now operating and likely to continue, which, fifty years hence, will produce marked results in the height, weight, and other physical properties of the system. In fact, we submit whether there are not already positive indications in the changed features, forms, and dimensions of the persons now on the stage, that a decline in this respect has actually commenced.

There is another topic which may afford us an instructive lesson, whether operating as *cause* or *effect* in the past, or at the present time. We have seen what a change has taken place in the disposition of men in respect to manual labor — that a strong disinclination prevails, especially among young men, to do farm-work, or to follow the more laborious of the mechanical trades, or any other kind of business requiring much hardship or exposure; and that, if possible, a greater change has occurred with women in respect to housework, as well as to all kinds of physical labor demanding great exertion or severe exercise. This aversion to manual labor or hard work, while it extends to all classes, comes, seemingly, more from young people than from the middle-aged. Now what mean these complaints, if a decline in physical vigor and strength has not already commenced? Besides, if such a course in respect to hard work is to be pursued by all our young people, what will be its effects ultimately upon their constitutions?

We come now to consider a different class of agents affecting the system, including food, drinks, medicines, etc., etc. And perhaps we can not introduce them better than by quoting the testimony of a distinguished foreign medical writer, who, after considerable observation and study, sums up the *"vices"* of the Americans under the following heads : —

1. An inordinate passion for riches.

2. Overwork of body and mind in the pursuit of business.

3. Undue hurry and excitement in all the affairs of life.

4. Intemperance in eating, drinking, and smoking.

5. A general disregard of the true laws of life and health.

Why our people should be so indifferent about human life, as such, or in the preservation of health, it is difficult to explain. Nevertheless, it is a fact. This indifference is strikingly manifest in their neglecting to take seasonable and proper care of themselves when ill; in trusting their lives, when sick, in the hands of empirics and charlatans; in taking large quantities of patent medicine, the composition of which they are entirely ignorant of; in swallowing compound mixtures highly recommended indeed, but by whom they know not; and the same trait is still more strikingly exhibited in a persistent, self-willed determination to continue certain habits or practices, which, from repeated warnings and expostulations, they know will hasten, if not cause, their death.

This disregard of life is also manifest in the general indifference of the public in cases frequently occurring, where one or more persons are drowned; in cases of suicide, or death by accident; and in those railroad disasters or steamboat explosions where human life is sacrificed on a still larger scale. How slight the shock, and how soon forgotten! Then, in the late war, what a sacrifice of life! What multitudes either killed in battle, or died by wounds, or by disease, and in prison under circumstances the most awful and appalling! Yet how soon will this terrible loss of life and all these heart-rending scenes be comparatively forgotten!

What this writer means by intemperance in "eating" is not so readily perceived; undoubtedly in many instances our tables are loaded with too much variety, as well as too great quantity of food; but it is impossible to adapt these two conditions to the habits, tastes, and health of all persons. We incline to think that this "intemperance" — at least a large part of it — consists in other things more peculiar to Americans, and that are decidedly vicious or injurious, among which may be enumerated:

1.   The hasty manner of eating — of bolting down food, without sufficient mastication, or giving time for the

glands in the mouth and throat to make the necessary secretions.

2. The practice of eating so much fine-flour bread, and that, too, frequently warm and poorly cooked.

3. The substitution of strong coffee and tea for plain or simpler drinks.

4. And the increased use of a rich, highly-seasoned, and stimulating diet.

Within forty or fifty years there has been a marked change in the mode of living — in some respects greatly improved, but in others calculated to impair the health and the constitution. The change in organization, to an increased nervous tissue, demands a change in regimen richer in quality, more highly seasoned, and stimulating. The appetite and taste both become more exquisite, more capricious, and exacting. For instance, tea and coffee are used not only in greater quantities, but must be made nicer and stronger, and the demand for condiments, as well as desserts, has greatly increased, and every year they must be made richer and more heating or stimulating.

At the same time, such a style of living and drinking serves to increase this nervous temperament more and more. The immediate tendency, therefore, of such a course, is to produce in this direction a species of physical degeneracy. It begets not only positive disease, but causes numerous weaknesses and complicated derangements in the system, which lay the foundation for complaints that may be partially relieved, but never cured. Such is neuralgia in all its endless forms, when based upon this organization. Every experienced physician knows full well the difficulty of treating or curing that class of diseases called "neuroses," and that they are constantly multiplying.

Under the head of "drinking," we include alcoholic

16

drinks, opiates, and tobacco. While there may be much plausible argument in favor of the moderate use of these articles in the type of organization found in the human system at the present day, still, according to all rational laws of physiology and pathology, their frequent, habitual, and extensive use must be condemmed. They should be resorted to only as *medicinal agents*, and then under the direction of wise and experienced physicians.

As these articles are now used in our country, there is no question but that they are decidely injurious to the constitutions of our people, vitiating their blood and under-mining their health, and producing more or less a species of physical degeneracy. They injure not only the body of the individual by poisoning his blood, by producing disease, and causing frequently premature death, but impair the sound, healthy action of the brain, thus striking a serious blow at mental habits, as well as moral and religious character. But the evil does not stop here. It is trans-mitted to offspring even to the third and fourth genera-tions, and sometimes in an intensified form. When the seeds of disease and vice are thus transmitted, it becomes doubly important to avoid the cause. That the continuous and excessive use of alcoholic drinks, as well as of tobacco, whether by chewing or smoking, causes fatal diseases of the liver, of the stomach, and of the heart, is admitted by the highest medical authorities. While no examination can show exactly how many are killed by either one or both of these agents, operating as first or secondary causes, it is undoubtedly true that great numbers do thus hasten their deaths. Besides, multitudes in the community carry through life the signs of decay from this source, in their walk, in their countenances, and throughout their whole physical systems.

Whatever differences of opinion may exist in the community as to whether the use of intoxicating drinks as a whole is increasing or not, there can be no question, we think, about the increased use of opium and tobacco. The sales of opium, in its different forms, have greatly increased within a few years, and it is well known to druggists and physicians that large quantities of this drug — mostly in the form of morphine — are consumed somewhat privately by individuals and families, either as an anodyne or as a stimulating agent. If the real facts could be obtained as to the fearful extent and the terrible effects of the drug as used in this way, it would surprise and alarm the public. When this drug once gets possession of its victim, there is no retreat, and the evidences of physical and mental degeneracy soon become well marked.

That the use of tobacco, particularly in smoking, is rapidly increasing, is too apparent to require proof. And what should cause most regret in the matter is, that so many young men are resorting to this pernicious practice. The evils of this habit have been so often pointed out in detail that it seems unnecessary to repeat them here, much less to dwell upon them. But we venture the statement that the use of tobacco, chiefly in smoking, is exercising a destructive influence upon the physical and mental energies of great multitudes of our people ; and that, by its continued increase, together with the law of hereditary descent, it is likely to result in an untold amount of physical degeneracy.

In respect to the other charges against Americans above mentioned, namely : "inordinate passion for riches, overwork of body and mind, and undue hurry and excitement," we admit there is too much truth in them, and that they have a most intimate connection with the subject under discussion. From remarks previously made on the nervous temperament, it will readily be seen that "hurry

and excitement " grow naturally in part out of the excess
of this organization, and at the same time they help to develop it more and more, thereby aggravating the evil.
This is one of the most discouraging features in the
matter as to any future reform, namely : that the very evil
itself operates both as a *cause and effect*, in conformity
with a well-known law of morbid action in the system.

The question might here arise, with free indulgence
of the appetite for strong drink, tobacco, and opiates, together with the strong love of riches, what may be the
effects of such habits in the development of the brain ?
Will they not tend to develop unduly those portions
where the vicious propensities and the selfish sentiments
have their seat ?   If so, what effect will this leave upon
the future habits and character of the people ?

But this "passion for riches," and this "overwork of
body and mind," have a broader, deeper, and far more
significant meaning than what appears upon a cursory
survey.   From this source spring some of the most powerful influences to undermine the constitution and the health
of our people.   We have set up a standard of living too
expensive — yea, extravagant — that has too many wants ;
we have not, as a people, the physical stamina or brain-
power to reach this standard and live by it, a few individuals may do it, but the many can not.   Multitudes,
attracted by the prize set before them, enter the arena, but
a few only reach the goal.

This standard is powerful in its influence, including
the fashions of the day, the equipage, the style, and the
manner of living, both of individuals and families.   Its
mainspring is *money — money — money —* which, in the
language of the wise man, "answereth all things."   Consequently, *money* must be obtained by all means, and at
whatever hazard and cost.   The appeal it makes to the
young, and particularly to heads of families, is almost

omnipotent, and reminds us of the famous lines of the classic satirist, which have been not inaptly rendered :

"My friend, get money; get a large estate,
By honest means; but get at any rate."

In this struggle for gain, what multitudes early break down in health, bring on disease, and sink into premature graves! How fast some men grow old, what peaked and haggard countenances, what careworn and wrinkled features, what frail and lean bodies, do we behold! What a large number of business men die in the *prime of life*, at thirty, forty, and fifty, when they should live to sixty, seventy, and eighty! How many *sudden deaths* occur among such men, which would not in a healthy state of business!

How rapidly changes the organization of young men, in this battle of life, from freshness of countenance, vigor of muscle, and elasticity of spirits, to the pale complexion, the feeble body, and languid gait! And the changes wrought in the constitution, health, and spirits of our women, in the headlong pursuit of fashion and style, are no less marked and deplorable. Do not such evidence and illustrations of change fully justify the conclusions we have drawn from them? Can any one foretell what is to be the result of this state of things? Does it not tend to establish the whole structure of society, as well as civilization, more and more upon a selfish and money basis?

The Human Body: Its Relations to Civilization.

THE first impression might lead one to suppose that such a subject as civilization would have little or no connection with the human system. But upon a careful investigation it will be found that the vital forces and the laws that govern the body sustain most intimate relations to true civilization. The very term *civilize* means to reclaim man from a savage state, to teach him the arts and all kinds of useful knowledge, to refine his manners, improve his habits, and secure for him the greatest possible amount of comfort and happiness.

Now, the most important agent and object in all these changes is man himself. His nature or the laws that govern every part of his body must be more or less affected by these changes. Whatever goes to make up civilization, or whatever changes are brought about by it, these should harmonize with the nature of man. There can be no permanent or true civilization, unless it is adapted to develop the whole nature of man. After a careful analysis of the elements of civilization — what it has been in the past, what it is at the present day, and what it should be — we find that its true foundation must be based upon the development of man's physical, mental, and moral nature, each in harmony with the other, and all to their highest extent. This grand idea or plan has, we believe, never been attempted, nor seriously considered, much less accomplished.

The Greeks and Romans made advances in this direction, in developing the body and cultivating the mind, but
failed in the moral element. We have at the present day
certain types of civilization which would be considered by
some superior to any in the past, and by others not easy
hereafter to be surpassed. But before we can determine
what is true civilization, or before there can be a general
agreement upon it, we must have some standard by which
t can be tested. This must be the highest possible development of man's whole nature, the animal and intellectual obeying the moral and religious. Then true merit
and real worth would receive their just reward.

Every organ in the body is governed by its own law, and
all parts of the system sustain certain relations to other
parts as well as to external objects. As the brain is the
crowning organ in the body, it is of the highest importance
that all its functions should be properly and harmoniously
exercised, and that to the fullest extent of which they are
capable in a normal state. Such a development would
constitute a normal state of physiology, and the central
point toward which all civilization should be directed.

Until some such state of society is brought about there
can be no true or permanent civilization. Its type must
be artificial, and to a great extent unnatural and unhealthy.
As long as it rests upon such supports, it must be unsatis-
fying and constantly changing. It partakes very much of
this character at the present day. A distinguished writer,
in characterizing it, maintains that wealth, fashion, and
show are its principal supports. This criticism might
at first seem unjust, but after all there is too much
truth in it. One thing is certain — there is no general
standard or agreement, and it would be difficult to decide
in what direction progress is leading us.

The fact is that in all the discussions on this subject
scarcely any reference is ever made to the body. The

most voluminous writer on civilization, Guizot, speaks of the domestic affections, the intellectual powers, and moral forces, but never discusses the relation which these classes sustain one to the other, or whether they have any connection with the body or the brain, or whether there can be any change in the physical system, or whether the organization of individuals or races makes any difference.

Almost the only writer who has attempted to apply physiological principles to illustrate changes in history and the state of society is the late Dr. John W. Draper, of New York. His work, "History of the Intellectual Development of Europe," is not only a monument of thought and research, but an honor to the medical profession. Says Doctor Draper: "Social advancement is as completely under the control of natural law as is bodily growth. The life of an individual is the miniature of the life of a nation." What we want is a practical application of these two propositions to the present state of society. It appeals emphatically to the members of this profession, as the body is peculiarly entrusted to their care and treatment.

It may be thought by some that these suggestions are visionary and unworthy of thoughtful consideration, but such will not be the verdict of posterity. In reviewing the history of physiology, who will assert that there can be no new discoveries in this science, or new application of its principles? What is the testimony of its teachers and professors — the highest living authority on the subject? Is it not that richer rewards await the votaries of this science, that the human family are to reap here golden harvests? History teaches that the great truths of Nature are slowly brought to light at different periods and by a variety of agencies.

Intermarriage of Relations.

THE intermarriage of relations is a subject which has always created much interest. The natural instincts of man seem to shrink from the idea of any such alliance between brothers and sisters, or parents and children, and also of unions in the same degree of relationship, as between uncles and nieces or aunts and nephews. But when we come to the third degree — that of cousins — there does not seem to be the same natural aversion, and such marriages have become, at the present day, quite common. Some writers have considered the parties entering into this union the same, whether related to each other by consanguinity or affinity — that is, by marriage. For instance, a man must not marry the sister of a deceased wife any sooner than his own sister, nor must a woman marry the brother of a deceased husband any more than her own brother. Others have taken a very different view of this last relation, and hence there have been many such marriages. Nearly all the leading religious denominations in Europe and in this country have condemned the intermarriage of persons related by affinity as well as those related by blood, and some of the most bitter personal controversies have arisen from this source that can be found in the whole history of the church. From a literal interpretation of the law of Moses, as well as a just construction of language, there are some grounds for this view. Most theories and legislation upon this subject may be traced directly to the law of Moses, as laid down in the eighteenth and twentieth chapters of Leviticus. From the

improper habits and practices of different nations and
tribes of people prior to this period, it is supposed that
it was found absolutely necessary to promulgate certain
laws or principles by which all such alliances should be
regulated. In the eighteenth chapter of Leviticus may be
found a very minute description of various relationships
wherein no such unions should ever be formed ; still, a
good deal of discussion and difference of opinion have
arisen upon the meaning of the phrase "near of kin,"
found in the sixth verse of that chapter.

While it has been very generally agreed that the law of
Moses extended only to the third degree of relationship,
portions of the Church have extended it even to the
seventh degree. Most civilized nations, that have legislated at all upon the subject, have based their statutes upon
their understanding of the law of Moses. So of the religious denominations, though some have been far more
rigid than others in the practical application of their interpretation of those laws. The Grecian and Latin fathers,
as well as the early reformers, adopted the Levitical law.
The Roman Catholic church early introduced it as a
canon ; and the Episcopal, together with the various
branches of the Lutheran, the Presbyterian, the Dutch
Reformed, etc., have considered it generally a fundamental
part of their creed. But of late years there has been
more and more of a growing laxness as to the observance
of these laws, as well as of discipline on the part of the
churches belonging to these various religious bodies. The
term *Incest* has been applied to the violation of this Levitical law, and seems to have been pretty well understood by
many of the nations of antiquity. Not only the Levites
and Jews generally understood it, but the Canaanites and
Egyptians, as well as the Greeks and the Romans.
Socrates said of incestuous marriage at Sparata and
Athens that they were " prejudical to healthy propagation

of the species "; and Plutarch, referring to the laws of the Romans on this subject, says : " They abstain from marrying their kinswomen in every degree of blood."

History shows, that among the rulers, generals, and eading characters of ancient Syria and Egypt, there was an unusual amount of such intermarrying, and that almost invariably it turned out badly. In the history of the royal founders of different nations there has been at times a great number of such intermarriages, and it is well known that very many of these have proved decidedly unfavorable as far as offspring is concerned. So marked was the effect that a peculiar disease, called the " king's evil," was said to originate from this source, and to become very common and troublesome. The power and influence which these families had endeavored in this way to perpetuate, has come to naught, and the name in history almost extinct.

In different nations, and at different periods, an immense number of facts have been collected showing the bad effects of the intermarriages of relations. These effects have been particularly marked in producing a large number of children who were deaf-and-dumb, blind, idiotic, feeble-minded, predisposed to insanity. The most striking illustration of this kind was collected by Dr. S. M. Bemus, of Kentucky, and published in the " Transactions of the American Medical Association," for 1858, which were as follows. Doctor Bemus, reported 833 such marriages, giving the time of marriage, the occupation, the temperament, the health, habits, etc., of the parents, with the number of children, their defects, peculiarities, etc., etc. The whole number of children was 3,942, of whom, 1,134 were defective: 145 deaf-and-dumb, 85 blind, 308 idiotic, 38 insane, 60 epileptic, 300 scrofulous, 98 deformed, and 833 died early. The proportion reported deaf-and-dumb, blind, idiotic, scrofulous, and deformed is altogether larger than what would be found among the children of families in the community, taking them indiscriminately. The degree of relationship in these cases is thus given : 10 marriages between brother and sister, or parent and child ; 12 between uncle and niece, or aunt and nephew ; 61 between blood-relations, who were themselves the descendants of blood-relations ; 27 between double-first cousins ; 600 between first cousins ; 120 between second cousins, and 13 between third cousins. In a careful examination of the several degrees of consanguinity here given, the hereditary effects are found to be the worst in the first and second degrees, in the third not so bad ; but when we come to the fourth, fifth, and sixth, the difference is not so perceptible.

As those cases were gathered largely at the West, in Kentucky and the adjoining states, we present an illustration from New England.

In a report presented to the legislature of Massachusetts, in 1848, by Dr. S. G. Howe, making inquiry as to the causes of blindness, idocy, etc., in the state, the following facts were elicited : The parentage of 359 idiots was ascertained. Seventeen were children of parents known to be near blood-relations. Three were from parents who were cousins. One-twentieth of all the idiots examined were offspring of blood-relations. The intermarriages of cousins do not constitute one-twentieth of all the marriages — probably not one five-hundredth ; therefore, the union of blood-relations produces more than its share of idiots. All sorts of human defects follow such marriages in a similar ratio, namely : deafness, blindness, insanity, rheumatism, excitability, etc. Of the seventeen families where the parents were blood-relations, most of them were intemperate or scrofulous. In some cases both conditions prevailed. There were born among them 95 children, 44 of whom were idiots ; 12 others were scrofulous or puny. In one family of 8, 5 were idiotic. Of the seventeen

families, the account stands thus : 6 have 1 idiotic child
each; 2 have 2; 3 have 3; 5 have 4; and 1 has
5 — making, of the 95 children from seventeen families,
44 idiots; that is, nearly one-half were imbecile.   Add
those otherwise imperfect, the 12 scrofulous and puny,
the 1 deaf, and the 1 dwarf, then we have 58, showing
more than one-half in an abnormal condition. The parents
who had the four idiots had also four other children who
were deformed.

A great amount of statistics on this subject has been
gathered by different writers, and in various localities,
which would certainly seem to indicate that some general law must lurk beneath them ; yet no very definite
principles or settled laws have hitherto been deduced
from them.   While new discoveries in science and improvements in society generally have been gradually clearing up many obscure and complicated questions, those
connected with the effect of such intermarriages still remain in a great measure unsolved — resting in a kind of
mystery.   That these important facts, so extensively and
rapidly accumulating, should thus stand out unexplained
is an opprobrium to physiology, and constitutes, as it were,
an exception to the results of inquiries in other departments of science after the true secrets or laws of Nature.

The great difficulty here has been the want of some
universal standard of appeal, or some general law of increase founded in Nature, whereby all these facts could be
properly classified, and then correctly tested   We propose
now to suggest such a standard or law, and apply it by
way of test to this great class of facts, and see whether
any new light can thus be obtained, by which these doubtful and disputed points may be settled.   If all facts and
theories on this subject can be harmonized — can be shown
to have a foundation in Nature, and can be reconciled by
application of the law about to be suggested — it will cer-

tainly afford very strong evidence of the existence of such a law.

Without going into details as respects all the minor points or the particular evidence in support of this law, we will here present its substance in as brief a manner as possible. It is *based upon a perfect development of all the organs in the human body, so that there shall be a perfect harmony in the performance of their respective functions.*

This standard of physiology, as represented by a complete development and harmony of the temperaments, is the same perfect organization of man as when he came from the hands of his Maker, and was pronounced not only "very good," but was commanded to be "fruitful and multiply and replenish the earth." But by man's course of disobedience and rebellion he lost not only the moral likeness and image of his Creator, but that harmony and perfection in his physical organism which he has never yet been able to regain.

In the discussion of this subject we have before us three central points or leading factors that sustain most intimate and important relations to each other. These are : 1, The facts connected with the intermarriages of consanguinity ; 2, The general principles established by Nature for human increase; and 3, The laws of hereditary descent. But before attempting to examine these relations in detail it may be well to notice particularly some facts in the history of the world, that led to the promulgation of the Levitical law upon this subject.

There is good reason to believe that, in the early history of the race, the intermarriage of kindred was a thing of frequent occurrence. From the time when the sons and daughters of Adam are supposed to have intermarried, before the laws of Moses, all the prohibitions then laid down upon the subject of marriage had been violated more or less in every age and nation. Those ancient patriarchs

and distinguished leaders among the people of God —
Abram, Isaac, and Jacob — all married wives connected
with them by consanguinity, and no intimation is made in
the Scriptures of the violation of any law in their case, or
that their progeny suffered any deterioration by such
alliances. But when the chosen tribes of Israel were
about to enter the land of Canaan, and be exposed to new
and manifold temptations, it pleased the Almighty to
promulgate some new laws for their protection and future
prosperity. With the exception of the Israelites, the
world at that time had become terribly corrupt and wicked,
particularly the inhabitants of Canaan. They had become
not only the worst kind of idolaters, but every vice that
could degrade human nature or pollute society was ram-
pant among them. " In the black catalogue of these," says
Professor Bush, "the abominations of lust stand pre-
ëminent ; and whether in the form of adultery, fornication,
incest, sodomy, or bestiality, they had now arisen to a
pitch of enormity which the forbearance of Heaven could
tolerate no longer, and of which a shuddering dread was to
be begotten in the minds of the people of the covenant.
And, in order that no possible plea of ignorance or un-
certainity might be left in their minds as to those connections which were lawful and which were forbidden, the
Most High proceeds to lay down a number of specific prohibitions on this subject, so framed as not only to include
the extra-nuptial pollutions which had prevailed among the
heathen, but also all those incestuous unions which were
inconsistent with the purity and sanctity of the marriage
relation."

It was in view of such vices, crimes, and exposures, that
we have promulgated those divine commands and prohibitions in the Levitical law as to the relations between
the sexes, accompained with their awful threatened penalties. But are we to suppose the revealed law the only

command ?   Had not those commands or prohibitions their
counterpart in human nature ? in the great laws of increase
— in the physical laws of life and health ?   Would a divine
lawgiver institute such specific commands to subjects, if,
as Creator, he did not know there had been created in
these very subjects an adaptation for these commands,
and that practically there must be harmony between the
natural and revealed law?

It will be readily admitted that such vices and crimes as
are here referred to can not be practised long without
generating the most loathsome and fatal diseases.   If only
now and then an individual was guilty of such sins the
consequences would not seem so serious; but when large
numbers, or a whole people, become thus vitiated and
corrupt, the effects are terrible.   We find abundant
evidence of this in the history and extinction of the
Canaanites ; as of other people given up to such practices.   Now, one of the preliminary steps or first stages
that led to such a demoralization in ancient times was undoubtedly the intermarriage of relations.   And the Almighty, knowing full well what was the nature of man —
what was necessary for the best protection, development,
and perpetuation of the race — saw fit to issue those rigid
instructions found in Leviticus.   It was not merely to
guard his people then from temptation, or from intermarrying among the heathen, or simply to promote good
morals among themselves, but to afford guides for the
formation of such alliances in all coming time.   And the
spirit, the intent of those instructions as a whole, are more
important than what their exact letter or literal interpretation, in the opinion of some, would seem to imply.
Neither should one of the minor or doubtful prohibitions
be unduly magnified, without properly considering its
practical application or particular adaptation.   Thus, in the
case of a man's marrying his wife's sister, some religious

denominations have considered it a criminal offense, and punish the offending party with the severest discipline of the church.

In order to a better understanding of these facts as connected with physical organization, let us recur to the laws of hereditary descent and human increase. Now, by applying the principle of hereditary descent that "like begets like," and that where two parties unite in marriage, of similar temperaments, with the same organs defective in structure or morbid in function, the evil in their offspring becomes intensified, if not doubled. It is well known that there are generally in families certain resemblances of features, form of the body, complexion of the skin, color of the hair, and frequently of physical and mental qualities throughout. Now, if marriage alliances take place with parties belonging to these family connections, and if these same individuals happen to have similar temperaments, with the same morbid tendencies to disease, what is the effect? Will it not transmit to the offspring imperfection of organization, weakness of the body, tendencies to disease, mental peculiarities, and all these increased, intensified? Let such alliances be continued through two or three generations, and what will be the effect but to make these evils worse, if not gradually to destroy the power of reproduction? Here comes in the great law of increase or population, which is the source—the fountain-head—whence all the laws of hereditary descent have their origin, their springs, their authority. It is that perfect, harmonious development of the system with which man was created, and which must always be our standard of imitation and appeal. It consists in the perfect development of all the organs in the body, with the harmonious, healthy performance of all their respective functions.

And the nearer this standard is approximated in married life, with all other conditions favorable, the greater

17

will be the number of children possessed of a full amount of vitality, vigor, strength, and health. While, on the other hand, the further we deviate from this standard, the more defective the organization of the offspring, and the less and less numerous, so that, when carried to the extreme point, the race is in danger of becoming extinct. For instance, let the physical organization become more and more defective, with a constantly-increasing amount of weakness, ill-health, and positive disease: there is not only marked degeneracy, but an inevitable tendency to the extinction of such a people. Or let the animal nature alone gain complete ascendency — its wants absorb all the attention of the mind as well as the labor of the hand — and the whole body sinks into a gross, sensual, and beastly state ; we shall find that Infinite Wisdom has set limits to the race in this direction. Or let the nervous temperament become altogether too predominant let it absorb all the nutrition of the body, let the mind ascend to the highest possible state of refinement, culture, and spirituality, to the neglect of the body as such : we shall find the offspring physically growing feebler, weaker, fewer in number, and finally incapable of propagating the race.

Occasionally we find in the writings of well-known medical men sentiments favorable to the views here advocated, though penned a long time since. Such is the following quotation from Doctor Pritchard : " Nature seems to have designed that the conditions and tendencies of human organisms should be kept very nearly in a state of equilibrium. This equipoise necessary to the healthy condition of man, upon whatever inexplicable cause it may depend, may be easily disarranged by giving undue predominance to any particular conditional phase. The slighter deviations from a normal mean would constitute individual or family peculiarities ; while more marked perversions become morbid manifestations, and infirmity results. As in the moral man none are exempt from the taint of sin, so in the physical man each individual of our race has his obliquity toward disease — generally, perhaps uniformly, toward some particular disease. It is, then, reasonable to expect that, when two individuals marry who possess the same morbid proclivity, their offspring will exhibit that identical divergence, but in a much more marked degree. Thus undoubtedly have originated many family peculiarities, perverted tastes, and morbid diatheses." The first two sentences in this quotation seem to express precisely the same condition or state of physiology upon which the law of human increase, as here advocated, is based. Had Docter Pritchard lived at the present day, the cause why this equilibrium or equipoise is so necessary to the healthy condition of man would not probably have appeared to him so "inexplicable."

If a careful and thorough investigaton should be made into the laws of population and hereditary descent, it will show, we believe, aside from a few isolated passages in the New Testament and the mandates of Moses, that the *marriage institution has a broad, sure, and unquestionea foundation in the laws of Nature itself.* It can be shown, by these *laws alone,* that for the healthy increase and perpetuation of the race, such an institution as that of marriage is absolutely necessary. And as a part and parcel of the conditions and results of this institution, comes in here this question of the intermarriage of relations. It is unnecessary to attempt to account for the ill effects of these marriages by advocating that there is some *"organic vitiation"* in such cases, or that there is something mysterious in the *"blood of kindred."* All the effects of such unions, however singular and conflicting, can be explained upon altogether more rational and satisfactory grounds. Admitting the fact that there is a greater resemblance, likeness, similarity, in family connections, extending sometimes to almost every organ in the body, than would be found in the same number of families not at all related, and that, when these connections form matrimonial alliances, it must have, according to the laws of hereditary descent, a marked and decided effect upon their offspring — if, in addition to this resemblance or likeness, these same parties should both have internal parts of the body imperfectly developed, morbid, or diseased, the effect must be still greater and more injurious. The nearer the relation, and more imperfect and diseased the bodies of both are, the effects become still more obvious as well as injurious.

No general rule of evil tendencies or bad effects from the intermarriage of consanguinity can well be established upon that fact alone ; but, upon certain conditions that are usually found connected with such alliances, certain effects can be predicted as most likely to follow. And inasmuch as these conditions, in the very nature of things, are more likely to be found existing in these cases, and are well known to prove more unfavorable than favorable to offspring ; and, moreover, as the evil tendencies rapidly accumulate with each successive generation, it was upon this ground that the Levitical law was established, and on this account such marriages should not take place.

There may be instances of such unions where the organization is so good on both sides, where the laws of life and health, of hygiene and good morals, are so well observed, that no bad effects whatever follow, and where the children and their descendants have the best health, reach the greatest longevity, and obtain distinguished positions and character in the world.

There are two phrases in common use, which, when applied to the laws of hereditary descent, are very expressive and full of meaning, viz. : "good stock," and "it runs in the blood." The primary signification of the word *stock* is stem, trunk ; their lineage, ancestry, race, etc. As the term is here used, it has a kind of collective or compound sense, meaning the highest and best qualities, physical and mental, combined in some one individual descended from a noble line of ancestry. Such examples may be found even in the line of consanguineous marriages, though more common where there is no relationship. By the word *"blood"* is meant, not that vital fluid common to all and essential to life, but some marked traits of character, or peculiarities in one family, lineage, or ancestry ; but, as more generally used, it designates in the individual some act or qualities that are mean, low-lived, vicious, sensual, characteristic, by association and inheritance, of the family or ancestry, etc.

There are two other phrases frequently used in reference to the improvement of domestic animals, and which are not devoid of meaning when applied to the human race. These expressions are, *"in-and-in breeding"* and *" the system of crossing" ;* and they derive all their meaning and force from the laws of increase and hereditary descent. For a definite and complete understanding of these phrases, the reader is referred to works treating of the improvement of domestic animals, where such principles can be far more readily and effectively applied. What has been found out here by a long series of experiments, could have been ascertained, in a great measure and far easier, by a knowledge of the laws here discussed.

Darwin, who has investigated biology more thoroughly than any other writer, makes the following significant remarks. Says he : " It is apparently a universal law of Nature that organic beings require an occasional cross with another individual." Again he says : " Nature tells us in the most emphatic manner that she abhors perpetual self-fertilization." And in another place he remarks that "marriages between relations are likewise in some way injurious ; that some unknown great good is derived from the union of individuals which have been kept distinct for several generations." But as to the philosophy or reasons, just "*in what way*" the evil following such marriages occurs, or can be avoided, or how or by what principles in different unions that "*unknown good*" can be obtained, Darwin does not tell us. The great beauty and value of such discussions depend very much upon *how far* they are made clear and intelligible to the common mind, and to *what extent* it is shown that they can be reduced practically to every-day life. Such a result is certainly very desirable in discussing a science like physiology, which professes to expound the laws that govern every human being.

The inquiry naturally arises, then, What are the relations which these laws practically sustain to the published facts upon this subject ? It was stated that the nearer the kindred or blood in such marriages, the worse were the effects. In the report by Doctor Bemus there were ten cases of marriages between brother and sister or parent and child, and twelve between uncle and aunt or aunt and nephew ; the former class had 31 children, 29 of whom were defective, and 19 idiotic ; the latter class had 53 children, of whom 40 were defective, and 3 idiotic. Says Doctor Bemus: "The increase and diminution of calamities to offspring correspond so closely with the increase and decrease of relationship, as to fix the conviction firmly in the mind that multiplication by the same blood by in-and-in marrying does incontestably lead in the aggregate to the physical and mental depravation of the offspring. And if we admit this statement, which the facts here abundantly prove — that defects of offspring multiply precisely as we multiply the same blood — and to this admission join the fact that all those contingent circumstances of parental health, habits, proclivities of constitution, sexual incongruity, etc., are as liable to affect one class as another, we can not rationally assign these effects to any other influence than consanguinity."

There are two classes of facts bearing on this point that should here be noticed. It was stated, when speaking of the law of population, that in proportion as the standard of physical organization is let down, in balance and quality, the less in the aggregate will be the number of offspring, and the greater the tendency to unproductiveness. Accordingly, in the cases reported of such marriages, it is found that the nearer the kindred and the poorer the organization, the fewer the children and the greater the number of *sterile* cases. As there may be other causes of sterility, it may be difficult to establish any fixed rule, but we think the principle here stated will, as a general thing, hold good.

There are two other correspondences found between the law and the facts. It was said, in describing the law of hereditary descent, that the more the parents were affected with scrofulous and other diseases, the more detrimental were the effects upon the offspring. An abundance of facts could be adduced in support of this position. And, on the other hand, the opposite doctrine is no less true, that where the organization of parents is good, the children generally will not suffer in their physical or mental qualities.

This doctrine of hereditary tendencies is clearly pointed out in the Bible, both by precept and example, in numerous instances. When Jehovah issued His commands in the decalogue, not only to the Israelites, but to His creatures in all coming time, saying, "*I the Lord thy God am a jealous God, visiting the iniquity of the fathers upon the children unto the third and fourth generations,*" it was intended that there should be *some meaning in this visitation.* Whatever divine influences or agencies may be brought

into operation in other respects, it is positively certain that
here, by laws of hereditary descent, the iniquities of the
fathers *are visited upon the children* unto the second,
third, and fourth generations. The term "iniquity" has
a broad signification, including the consequences or penalties of violated law, whether that law be expressed in
the revealed command of God, or stamped by the same
Almighty power upon the human constitution.

As there are only two classes of marriages in question
that really, at the present day, create very much interest
— that is, the intermarrying of cousins, and the marrying
of a wife's sister — it may be expected that the merits
and demerits of these should be particularly noticed. In
respect to the former class, the objections and prejudices
which once existed against such unions seem to be dying
out, and such marriages are becoming more and more
common.

It is a grave question still with some divines, whether
it was the intention or not of the Levitical law to extend
to cases of this latter kind. The relationship by affinity,
making the wife's sister the same as a sister by blood, does
not seem to commend itself generally to the common-sense
of mankind.

The fact that she has, by chance, been a member of
the same family or in frequent social intercourse with
it, in the opinion of some would present an argument
against the formation of such alliances; on the other
hand this very experience and acquaintance — especially
where there are small children — would constitute a
very strong argument in its favor. If the question
was to be settled by any principle, it should be on the
ground of organization and adaptation of the parties interested. The effects of such a union are far more important and permanent in this direction than anywhere-
else; and it was mainly to reach and regulate these, that

the Levitical law was first established. The strong
tendencies of the popular mind at the present day — and
we think physiological laws sustain this sentiment — are in
favor of such unions.

But in respect to the intermarriage of cousins, it has
never been asserted that these were specifically interdicted
by the law of Moses, though very strong prejudices have
always existed in the community against the practice.
We can not, perhaps, express the opinions that some persons entertain upon this subject better than by quoting
the following sentences from one of our popular periodi-
cals. Says this writer: "Whether cousins should intermarry, or be considered as within the forbidden degree, it
is indisputable that many cousins have intermarried; and
it is equally beyond dispute that, in many instances, the
offspring of such marriages are, to all appearance, as well
calculated to make good figures in life as any young persons. Every man must have among his acquaintances
families like those spoken of and can testify that the marriages of cousins do not always lead to the birth of idiots.
Hence it is not just to speak of such marriages as if they
must lead to the increase of imbecility; and the sweeping
language that is sometimes employed on the subject, besides being unsupported by facts, causes no little misery
to men and women who do not deserve to have their lives
made wretched by extravagant assertions that would place
them in the black list for life. Cousins who have married,
and to whom it never occurred that in marrying they violated any moral or physical law, are startled when they
read in respectable publications that their children's only
inheritance is to 'all the ills that flesh is heir to.'"

But inasmuch as there is in cousins, according to the
best statistics gathered, so much of a family resemblance
or likeness, and so much imperfection of structure, or
morbidness of function, or eccentricity of character is

found to accompany it, that the hereditary effects are pre_
judicial to offspring, we are warranted in laying it down
as a general rule, *that cousins should not intermarry unless
possessing sound constitutions.* This rule is the more imperative, when it is considered that the evil effects accumulate with wonderful power in every successive generation,
and that there is no absolute necessity for the formation
of such marriages. *As free moral agents and accountable
beings, then, we have no right to inflict upon the innocent
such an untold amount of misery and suffering.*

At the same time, there may be found many instances
of such intermarrying, where the parties have good organizations and are more healthy than the average run of
people, with children equally as numerous and healthy. In
such cases there can be, as far as we can discover, no valid
objection whatever to cousins intermarrying. This representation of the case is in accordance with the great body
of statistics collected, is sustained by the laws of human
ncrease and hereditary descent as here set forth, and will
commend itself, we doubt not, to the common-sense and
intelligence of the most enlightened portions of the community. It places the responsibility, whether for good or
evil, upon the free-agency, knowledge, and moral sense of
the individuals most interested, — just where God designed
it should be.

One object of discussing this question of intermarriage
of relations was to show that the Levitical law had its
counterpart in the laws of physiology. There must have
been evils of the most flagrant character, from such
unions among the Israelites, the Amorites, the Canaanites,
etc., that led Moses, the lawgiver of Israel, to institute
those laws found in Leviticus. And history shows that
these specific directions in respect to the intermarriage of
relations had a foundation in the laws of the *physical
system* — that they were no mere arbitrary commands, a

dead letter in the decalogue, — but were sustained by the principles of physiological science. Whatever effect the promulgation of these laws had at the time upon the Israelites, it is evident that some knowledge of them extended to heathen nations and commanded their respect. They had a powerful influence among some of the most enlightened people of antiquity ; and wherever civilized nations have legislated at all upon marriage, these laws have been the basis.

Among the largest religious denominations in the world these laws have had a great influence, and, in some of them, they have been adopted as a canon in their creed and practice. The Mohammedans, notwithstanding polygamy is practised among them, are shocked at such intermarriages. The Koran has this remarkable passage : " Ye are forbidden to marry your mothers and your daughters, and your sisters and your aunts, both on the father's and on the mother's side ; and your brother's daughters and your sister's daughters, and your foster sisters and your wives' mothers, and your daughters-in-law who are under tuition, and the wives of your sons ; and ye are also forbidden to take to wife two sisters." The Greek and Romish churches embody the substance of these laws in a part of their canons, and have been somewhat rigid in carrying them into practice, especially the latter body. The leading branches of the Protestant church — the Episcopal, the Presbyterian, the Baptist, etc. — have adopted these laws as a part of their creeds.

The human body was the same, and was governed by the same laws, in the time of Moses as at the present day. The Levitical laws and the comments in the decalogue were wisely adapted to certain physiological conditions of the body. "Visiting the iniquity of the fathers upon the children unto second, third, and fourth generations," are the mere penalties attached to violated laws. There must be perfect harmony between the revealed commands of God and His laws stamped upon the human system ; and when correctly interpreted, there can not be any disagreement. The laws of hereditary descent are unmistakably taught in the Scriptures, but these can not be understood or applied, without having their origin and foundation in a general law of propagation. And we do not see how this whole subject can be satisfactorily explained without the admission of such a law. The question may be asked, why a law so important should not have been discovered before? Many answers might be given.

It seems to have been the design of Providence that the great truths of Nature should be slowly brought to light at different periods, and sometimes in a very incidental manner, as well as by humble agencies. The laws of gravitation were the same, and similar phenomena had been witnessed by multitudes, long before Newton, from the falling of an apple, caught the idea that there was some peculiar or occult power indicated by that apparent accident. The heart had been repeatedly examined by anatomists — its structure and functions had been carefully studied by large numbers of physicians — before Harvey discovered the circulation of the blood. So, the morbid structure and functions of the lungs had been made a special study by many medical men, long before Laennec discovered, by auscultation and percussion, that the physical outward signs could give such a wonderful insight into the pathology of this organ. All great truths, when once discovered, are very simple, and the surprise to all is, that they were not generally known before.

If the theory here advanced is the true law of human increase or population, it is not a mere theory or an abstract general principle, but it is capable of endless applications ; for instance, in affording us a better knowledge of man — his duties and responsibilities in relation to himself, to the family, to society, and to his Maker; in furnishing a guide, or great principle, by which certain practices and fashions in society, certain modes of education, systems of morals, legislation, etc., can be tested; in showing the importance and sacreduess of the laws of life and health, that they are a part of the will and government of God in this world, as much as His revealed commands.

In the discussion reference has been made to the harmony between the facts in science and the principles of Revelation. Thus, when certain discoveries were made in astronomy, they were at first thought to conflict with the Bible; the same was true in geology; but, by more thorough researches, a surprising harmony was found to exist between the teachings of Revelation and the laws of Nature. May not the same result prove true in physiology? As far as most practical purposes are concerned, we have as yet only reached the threshold — the vestibule of this temple of *the science of man* — with which, in point of actual value and utility, the sublime truths of astronomy and the more wonderful revelations of geology sink into comparative insignificance.

College Sports.

NEARLY twenty years ago, we made this statement :
"No system of education can be complete or carried
to the highest extent of which the mind is capable
while the laws that govern the body and the brain are
ignored or violated. The time will come, too, when the
officers of our colleges and universities will feel compelled
to take into more careful consideration the *physical wel-*
fare of students. And we venture the prediction that in
no department of education will there be greater improvement, for the next fifty years, than in the better care and
development of the human body." Though not half the
time here specified has passed, there has been a great increase of interest in physical exercises of some kind in
many of our highest institutions of learning. This interest has been manifested particularly in the direction of
gymnastics, and the building of gymnasiums. Within a few
years nearly fifty literary institutions in the United States
have either built gymnasiums, or signified the intention of
so doing and introducing physical exercises of some kind.
This is a movement in the right direction, and can not
fail in meeting with success.

There has been a great increase of interest in other
exercises of this kind, particularly in base-ball playing,
boating, and in the exhibition of "Field days." The
term "sport" is very appropriately applied to some of
these exercises, as in order to reap the full benefit from
them, there must be combined more or less sport and
amusement. Walking, the oldest of all these exercises,

and once extensively practised, is falling behind, but is one of the best. It calls into play most of the muscles of the body; requires no expense; can be pursued in the open air, and when one can have pleasant company and attractive scenery, no better or more wholesome exercise can be found. There is another kind of exercise, having some advantages over walking, that is horse-back riding, but this can not be enjoyed without incurring expense, and only in pleasant weather. One objection to both these exercises is that large numbers can not conveniently engage in them at the same time. It seems necessary there should be a crowd or a large company present in order to create sport and perpetuate the interest. The exercise must be of that kind which makes a display — can be seen and known to the public.

There is a game that creates great interest at times — that is, foot-ball. This is very exciting and is peculiarly calculated to develop the muscles of the lowest extremities. This game comes in very opportunely for a change or variety, and when properly conducted, it affords very wholesome exercise. But there are objections to this game, inasmuch as the movements are apt to become very violent, and cause some sprains to the legs or injury to other parts of the body. Besides, such is the haste or zeal in pursuit after the ball, that it does not get all the hits, but not unfrequently individuals are severely injured by kicks aimed at the ball, or by violent contusions of the limbs. Some attempts have been made to introduce into our literary institutions the military drill, or tactics as regular exercises ; but, on account of their monotony and the difficulty of infusing into them, for any length of time, a proper degree of enthusiasm, they have failed. If the sound of the fife and the drum could be heard, with the mutterings of war at a distance, there would be no lack of life and enthusiasm. Such was the case in several of our

institutions at the time of the civil war. In military
schools where young men are trained for government service, this drill, in all its variety of forms, has answered
an excellent purpose for improving the constitution and
promoting physical health.

Ball-playing and boating are peculiarly calculated to
attract attention and excite general interest. These
are well known, under the head of collegiate sports —
sometimes one claiming most attention, again the other.
Ball-playing has been so popular everywhere that it has
been denominated the "National game." While the
colleges may take the lead in this sport, a large number
of ball clubs are organized in our cities and larger towns.
If the college clubs were confined in the game with
others organized within an institution, the whole matter
could be managed much better.

If the merits of ball-playing and boating are to be
tested by the laws of health and sanitation alone, serious
objections are found to them when the game is played
under great excitement; there is danger, in the long-con-
tinued running and violent batting, of producing congestion of the lungs or an unnatural strain of the heart,
which might prove fatal or result in serious disease.
If the person has a constitutional weakness here, or any
abnormal formation, there is still greater danger.

Again: In ball-playing, a serious evil arises from the
fact, that a few and the same muscles are called into
exercise continuously. Then these portions of the limbs
and of the body where those muscles are located, require
more nourishment comparatively and become larger in
size and much stronger. This violates one of the first
principles in physiology for good health and long life.
Sanitation demands a harmony or balance in the organization. One of the greatest safeguards against sickness and
disease is a well-balanced constitution, and one of the

primary objects of gymnastics is to develop the weak parts, in order to equalize the strength. The exercise of ball-playing, taken moderately and pursued under proper restrictions, possesses superior advantages. In a simple form, it requires no great preparation or expense. It is not necessary to hire persons to perform the most difficult part of the work, nor that wagers or bets should be offered by visitors. The game can be played without having great crowds in attendance, or an excited public awaiting the issue. It is the abuse of a *good thing* that does the injury.

It is true, ball-playing has this special advantage, — it creates great interest in games and sports, and enlists heartily large numbers in physical exercises, some of whom have not been particularly before engaged in them. When this game is played so extensively in the country, with so much excitement, and the press filled with reports, the attention of the public is specially directed to physical development — to the importance of a sound constitution and good health. So far as waking up an interest in physical exercises or collegiate sports, this general ball-playing is a grand thing ; but the evils may counter-balance all the advantages. It is a question whether the only safe course for inter-collegiate sports is, that they be placed under a committee of the faculty of every college. As to gymnastics, these are placed under the care of the teacher of these exercises, where they are in safe-keeping. The exhibitions of field-day also should come under his supervision, as well as every exercise under the gymnastic department of the college.

18

Mental Philosophy:
Its Connection with Medicine.*

THE celebrated Doctor Rush, in enumerating the causes that retarded the progress of medicine, adduced as one of the principal, the neglect of cultivating those branches of science which are most intimately connected with medicine. These are, says he, chiefly "natural history and metaphysics." The former term he used in its widest sense, comprising both the animal and vegetable kingdoms; but by the term metaphysics he intended to include only that field of inquiry which relates to a knowledge of the operations and faculties of the human mind.

Though the above remark of Doctor Rush was made nearly a half-century since, yet it may apply, if we mistake not, with equal force and propriety to the present state of medical science. While every other branch of knowledge connected with medicine has been rapidly progressing, that styled here metaphysics has, to a very great extent, been treated with entire neglect by a large majority of this profession. Perhaps it may be safely stated, that in no other department of human improvement has there been a greater advancement for the last fifty years than in that of medicine. Every year has witnessed some important development of new truths, as well as a more correct application of those already discovered. Anatomy,

---

* This essay was presented by the writer to the Faculty of Pennsylvania Medical College, March 1, 1841, as an Inaugural Thesis for the degree of Doctor of Medicine.

Physiology, and Surgery, have each within this period been enriched by many splendid discoveries and improvements. Pathology, which then was scarcely known or recognized as a distinct branch of medical study, has since received special attention and shed a vast amount of light upon the causes, symptoms, and treatment of disease. The departments of Materia Medica and Therapeutics have also been greatly improved by many new discoveries in chemistry and pharmacy. Add to these the experiences and observations of many able and skillful physicians, and we have medicine in its present highly cultivated and improved state. But the same cause which Doctor Rush mentions as retarding the progress of this noble science, still exists. While every other branch of medical knowledge has been constantly advancing, a knowledge of *mind*, as far as medicine is concerned, has remained almost stationary for centuries. Dr. Southwood Smith very correctly observes that "the degree in which the science of mind is neglected in our age and country—and may it not be justly added, in our profession?—is truly deplorable." There must be some cause or reason for this state of things, and the writer proposes in the present essay to inquire —

1. *Why* the cultivation of metaphysics is so generally neglected by medical men ; and

2. To point out the intimate connection of mental philosophy with medicine ; and

3. To offer some remarks upon the importance of a knowledge of this science to the physician.

In the first place, it can not be adduced as a reason why mental science is not more successfully cultivated, that not sufficient talent, learning, and research have been devoted to the subject. Some of the best minds that the world ever produced have labored assiduously in this field of study, and their productions bear the stamp of

unwearied industry and profound attainments. Again :
This neglect can not be accounted for by any reason
deduced from the nature and unimportance of the subject.
All writers on the philosophy of mind have borne their
united testimony that a knowledge of the principles and
applications of this science is of the highest possible
importance. Yet the great and most efficient cause of
this neglect, as we apprehend, remains to be stated — it is
the *erroneous mode of investigation* that has been hitherto
employed ; the leading defects of which may be summed
up under the following heads.

First, Metaphysicians have omitted in their investigations almost entirely the intimate and necessary connection
that exists between the mind and the body. In all their
researches they have viewed the mind as an *abstract
essence* — as existing, and performing all its operations,
independent of any material instrument or agency. They
have treated not only with neglect, but with disrespect,
that great law established by an all-wise Creator — viz., that
*mind, in this world, should be dependent on physical organi-
zation for its manifestations.* This law constitutes the
only true foundation upon which any correct system of
mental philosophy can possibly be based ; and the consequence of overlooking this condition has proved most
disastrous on the cultivation of metaphysics. The very
term itself has become a by-word, and those who are
devoted to its pursuit are not unfrequently made the subject of remark and ridicule. To call a man (observes a
popular writer) a metaphysician, at the present day, is a
delicate mode of recommending him to a lunatic asylum ;
and Doctor Armstrong, the well-known writer on medicine,
has wittily defined metaphysics to be " the art of talking
grave nonsense upon subjects beyond the reach of the
human understanding." Doctor Bartlett, one of our
countrymen, very justly remarks, that " almost the whole

history of metaphysics is a record of absurdities, inconsistencies, and contradictions. The very name has become, almost by common consent, only another name for intellectual harlequinism and jugglery. Never has the human mind been guilty of playing more fantastic tricks than when attempting, by misdirected and impotent efforts, to unriddle the mystery of its own constitution."

Secondly, Writers on this subject have not only based their systems of philosophy on *reflection* and *consciousness* in general, but have erected their own individual consciousness into a universal standard. Says Dugald Stewart, in his philosophical essays, " All our knowledge of the *human mind* rests ultimately on facts for which we have the evidence of consciousness. And accordingly, in my inquiries, I have aimed at nothing more than to ascertain the laws of our constitution, *as far as they can be discovered by attention to the subjects of our consciousness.*"

This remark of Stewart will apply to nearly all the writers of the metaphysical school. But instead of consciousness being a true guide in mental investigations, it is decidedly unsafe and erroneous. In the first place, consciousness affords no positive evidence of the existence and functions of the cerebral organs, by means of which alone the mind acts in this life. It simply takes cognizance of mental operations in general, and throws comparatively but little light on the nature or number of the distinct faculties of the mind. Again : It is impossible for an individual to base evidence on this source alone, without considering *his own* consciousness as a standard for *all others*. This constitutes one of the most radical errors of the metaphysicians. They have taken their own minds as a standard, or type, for the whole human race ; and, accordingly, each has begun to erect a system or theory of his own, by demolishing that of his predecessor.

Now, the consciousness of no two persons is alike, any more than the features of their bodies ; and it is utterly erroneous, as well as absurd, to consider such a guide or rule as susceptible of universal application. The great variety of systems, theories, and speculations in mental philosophy have arisen, in no small degree, from this source. Hence, too, the great diversity of opinions, as well as contradictions in conclusions, on the part of those devoted to its pursuit. This very fact affords *prima facie* evidence that their premises were false ; and consequently, that their systems were not founded in Nature, whose laws, when correctly interpreted, are always harmonious and everywhere the same. Truth, like its Author, is ever consistent with itself.

Thirdly, Another radical defect in past methods of investigating mental phenomena consists in an almost entire reversal of the true mode of studying Nature. *Observation* and *experiment* are the only sources from which we can derive any positive evidence for the establishment of principles in science. Facts must first be observed and properly classified ; and when a sufficient number have been collected, or none of a contradictory nature can be found, general principles may safely be deduced from these, and be considered as permanently established. But instead of pursuing this slow and tedious process, as marked out by the immortal founder of the inductive philosophy, metaphysicians have first commenced by forming visionary hypotheses and assuming certain premises, and afterwards have attempted to reconcile facts with these. They have retired to their cloisters, and speculated by the light of their own consciousness, when they should have studied by observation and experiment the great book of Nature. They have capriciously allotted faculties to man, and arbitrarily dictated laws to Nature ; and the consequence is, there has been but little of truth mingled in their researches. Some have denied to the mind all *innateness* of disposition or character, and have maintained that it was precisely like a piece of white paper (*tabula rasa*), capable only of being acted upon, and moulded by, outward impressions. Others have assumed that all minds were by nature *alike* as to capacity, and that the great diversity in the talents of different men was solely occasioned by external circumstances. In fact, no two leading metaphysical writers can be found who agree as to the nature or number of the faculties of the human mind.

Fourthly, Another serious defect in past investigations on this subject is a complete failure to account for many mental phenomena. It is to be presumed of every true science, that it will afford some rational explanation of the principal causes and relations of the various phenomena of which it treats. But it is far otherwise with the one under discussion. Many facts in this science, as far as the labors of metaphysicians are concerned, now for more than two thousand years, remain to this day entirely inexplicable. They afford no rational explanation whatever of the following topics : *nature of genius; causes of diversity in talent and moral feeling among different individuals ; effects on the mind of opium, and other intoxicating substances, taken into the stomach; difference between the sexes ; the process of gradual development of the mental faculties ; the causes of idiocy ; the phenomena of dreaming, somnambulism, insanity, monomania, etc.* Moreover, the philosophy of the *will*, the laws of *free agency*, and the different degrees of *human accountability* have never yet been satisfactorily expounded by any system of metaphysics. Other instances of failure might be adduced, but certainly the facts and phenomena already mentioned, among the most important in life, should be clearly and rationally accounted for by a system of mental philosophy.

Again : It is fair to infer that a science which should give a correct exposition of the faculties of the mind, and the laws which govern their development, would be fraught with the highest practical benefit to mankind. But when examined by such a test, how directly the reverse of this are all the labors of metaphysicians! Their researches have been altogether too speculative and ethereal to be reduced to any practical purposes. The subject itself has not only fallen into disrepute, but, as a branch of study, receives scarcely any attention at the present day, in our seminaries and institutions of learning. In view of these facts, it is not surprising that the study of mental philosophy should have been neglected by medical men. Its principles, as hitherto taught, have had too little to do with *physical organization*, in order to come under their cognizance. But when the true mode of investigating the subject is correctly understood and admitted, it must devolve on the members of this profession to *take the lead* in its cultivation ; and they can then no longer continue to neglect it without violating the most sacred duties which they owe to medicine, as well as sacrificing the best interests of the public. This brings us to a consideration of our second general head.

2. *The Connection of Mental Science with Medicine.* — Before entering directly upon an examination of this question, it will be necessary to decide, or settle in some measure, what are *true* principles of mental science. It will be seen, from the preceding observations, that we can not rely upon the mode of investigation adopted by metaphysicians, neither can we obtain from this source a correct knowledge of the philosophy of the human mind. This fact must be admitted, we think, by all candid and competent judges. What, then, is the true foundation of mental science ? What are its principles, and the nature and amount of evidence in support of them ?

First, then, we have no positive knowledge whatever of mind as an abstract essence or entity. Though we believe it to be of an immaterial and spiritual nature, destined to immortality, yet God has never endowed us with faculties capable of comprehending or taking cognizance of any such existence. It is therefore useless to indulge in any speculations about its *nature* or *essence*, and folly to predicate a system of mental philosophy upon such a basis. All we can possibly know of mind, as manifested in this world, is through its material instrument. That the brain is the organ of the mind has been the united testimony of the best writers on Anatomy for centuries, and is, moreover, confirmed by the opinions of the highest living authorities on the subject. Here, then, is the first principle — the foundation of mental science. In the second place, the brain is composed of a congeries of organs, corresponding in number to the faculties of the mind. This is proved by analogy, observation, and experiment. The brain, as its anatomy shows on dissection, is a complex viscus or body, and is made up of distinct parts or organs. Now, according to a law pervading all organic matter, where distinct organs are found, however similar in structure, or nearly connected in their relations, they perform entirely different functions. The brain can not be an exception to this universal law. Again: The mind consists of a plurality of faculties, and, in accordance with the counterpart of the law just stated, it must necessarily have a plurality of instruments. And both observation as well as experiment prove that these instruments are distinct organs in the brain. Thousands, who have made accurate and extensive observations, and whose testimony can not be called in question, agree on this point. They have, moreover, collected such a number of facts in confirmation of it as to afford positive and irresistible evidence of its truth to every unprejudiced and well-disciplined mind.

It has also been found, by actual experiment in a multitude of instances, that whenever particular parts or organs
of the brain suffer serious injury, the corresponding
faculties of the mind have invariably been more or less
impaired in their manifestations.   No person can candidly
and thoroughly investigate this proposition without being
absolutely compelled to admit its truth.

The third great principle in this science may be thus
stated—the size of the organ, other things being equal, is
a measure of the power of its corresponding faculty. This
law is also one of general application.   The conditions involved in the phrase, "other things being equal," will of
course vary in character under different circumstances;
but when properly considered, size is strictly a measure of
power, and there can not be found an exception to the law
throughout the universe.   It is unnecessary here to accumulate facts either for the purpose of illustrating or proving this principle.

Our next inquiry is, Can we ascertain accurately the size
of these several organs in the living head? And secondly,
Can we, by making proper allowances for the influence of
these other conditions on size, judge correctly of the
strength of the different faculties of the mind? These
qustions must be settled by matters of fact and actual experiment.   They afford no chance for speculation or
sophistry, and none but those who have carefully examined
the subject are qualified to give testimony in the decision.
First, then, can the size of the brain and its various parts
be ascertained ? Says Magendie, "The only way of estimating the volume of the brain in a living person, is to
measure the dimensions of the skull." Sir Charles Bell
also observes, that "the bones of the head are moulded to
the brain, and the peculiar shapes of the bones of the
head are determined by the original peculiarity in the
shape of the brain." Blumenbach, Cuvier, Monroe, and

other distinguished anatomists have expressed similar sentiments. Thus by various measurements of the skull, then, externally, we can ascertain the size of the different organs of the brain. It is true, there may be certain exceptions to this principle, as in the case of disease or old age, but these by no means invalidate its truth, or the practicability of its application. Some difficulty may also occasionally be experienced from the extreme thickness or irregularity of certain parts of the cranium, but the precise nature or amount of this difficulty can generally be understood, proper allowance can be made for it, and very correct inferences drawn as to cerebral development.

Being able, then, to ascertain the size of the several organs of the brain, can we judge correctly of those conditions which influence or modify its functions? These are, chiefly, constitution, temperament, health, and education, a knowledge of which may certainly be ascertained, both from the organization of an individual, as well as from his own statements concerning his history and circumstances. This remark is not mere assertion — it is supported by a multitude of facts, and did the occasion require, we might furnish an amount of evidence in confirmation of its truth that could be neither disputed nor denied ; but, for the present, we must content ourselves by referring the reader to such works as treat particularly of those points. It may be asked, if we consider the above propositions sufficiently proved and established to be regarded as settled principles in mental science ? We reply in the affirmative. This question is not to be decided by our individual knowledge on the subject, nor by the ignorance of the community generally. All the great principles in physical science have been discovered, proved, and established by a few original minds ; and the truth of such discoveries is always to be admitted, not by the extent to which they have been propagated, nor by the mere number who publicly advocate them, but from the positive evidence furnished by their original discoverers and expounders. It is thus we judge in relation to the truths of chemistry, geology, and natural philosophy; and it is unfair, as well as unjust, not to apply the same rule to mental science. For its principles are based on precisely the same kind of evidence, appealing directly to the senses, observation, and experiment ; and we venture to hazard the opinion that its leading advocates are as competent judges in this matter as the teachers of any other great department of science.

Thirdly, It is well known that the state of the mind has a powerful influence over the body, especially when in a morbid or diseased condition. In no instance is this reciprocal influence more powerful, either for good or for ill, than in severe and unexpected injuries ; and under no circumstances whatever can it be brought to bear more efficiently than in surgical operations, which are attended with great difficulty and danger. There are undoubtedly many cases where the success of an operation, as well as the life of the patient, depends almost entirely on the state of mind or feelings at the time and afterwards. Now, a system of mental philosophy, based upon the functions of the brain, will afford the most essential aid in such cases. It will enable the surgeon to detect at once the strong and weak faculties of his patient, and thus assist in presenting such motives, and just such appeals, as will operate most beneficially on the feeling and spirits. Says Doctor Rush, speaking in relation to medicine in general, "The advantage to be derived from this source (i. e., a knowledge of mind) might be a hundred times greater, were they properly directed by well-educated physicians."

Pathology is comparatively a new light in medical science, as but little attention was given to the subject till within a few years. Its object is to investigate the

changes which have taken place in the functional derangement or structure of an organic body, either as the cause
or effect of disease. This mode of investigation has been
prosecuted with great zeal, talent, and industry, by many
of the most distinguished men in the profession, and it is
to this source, more than to any other, that we are recently
indebted for some of the most valuable discoveries and
improvements in medicine. Among other inquiries, the
morbid conditions of the brain have by no means escaped
the notice of pathologists. At the same time, we venture
to affirm that there is not another organ in the human
system which has received an equal amount of attention
as to its pathology, but what has been attended with more
definite and satisfactory results. The cause of this arises
from three sources : — viz., first, from the extremely delicate
texture of the brain ; secondly, from the very complicated
structure and intimate relations of its several parts ; and,
thirdly, from our imperfect knowledge of its functions.
The last, as we apprehend, is by far the most fruitful
source of difficulty in itself, besides being, to a considerable extent, the occasion of the two former.

Pathology, as a science, is based on physiology. For
an examination into the causes and effects of disease,
whether it be functional derangement or change in organization, presupposes necessarily a knowledge of the healthy
state and function of an organ. Otherwise we could not
judge accurately of the deviations from health, neither
could we understand the changes which have been occasioned by disease. And never can pathological researches,
as to the brain, be carried out and perfected, till the physiology of all its parts is thoroughly comprehended. This
knowledge is indispensable, in order to make proper
observations, and to establish general principles in pathology. First, if we were perfectly acquainted with the
functions of every distinct portion of the brain we should

then know precisely what parts to examine in case of disease, and should thus be far more likely to discover the morbid derangements in function, or the nice changes in structure, that may exist. Secondly, the various parts of the brain sustain very intimate and important relations to each other in the performance of their functions, including muscular motion, sensation, and mental operations. Now, these several relations and connections must first be understood in a *healthy* state, before we clearly perceive the causes or effects of diseases in all parts of such a complicated viscus. Thirdly, the brain is subject to a great variety of affections, where no indications or traces of change in organization have ever yet been discovered by the best pathologists. Whether this difficulty arises from the extreme delicacy of its texture, or the want of more perfect instruments for making the examination, it is unnecessary here to decide. But it frequently happens, as is rendered evident by external symptoms, that very great functional derangement actually exists, and according to all analogy, there is every reason to believe that some change in physical structure must either have preceded, or been occasioned by, this derangement. Now, a thorough knowledge of the functions of the brain, embracing the various kinds of motion and sensation, as well as mental manifestations, will not only incite, but enable us to recognize far more accurately, the *kind* and *degree* of deviations in these, from a state of health. We may thus, by continuing this mode of inquiry and examination, be able to detect changes in organization which have hitherto entirely escaped the closest scrutiny of pathologists. Hence we see that a knowledge of physiology must precede that of pathology, and that mental philosophy sustains, in this respect, also a most intimate and important relation to medicine.

Practice of Medicine.—Aside from good natural abilities,

two things are indispensably requisite to constitute any individual a successful practitioner of medicine. First, he must be thoroughly and practically acquainted with the causes and symptoms of disease; and, secondly, with the nature and application of the most appropriate remedies. And the more complicated the disease and difficult its treatment, the more important that his knowledge should be accurate, extensive, and well grounded. This is emphatically true, in reference to nervous diseases. It is stated in the Library of Practical Medicine — the most recent and popular work on the subject — that "the diseases of the brain are, at the present moment, more obscure than any other great class in the nosology."

While there has been a constant improvement in the diagnosis and treatment of diseases affecting every other part of the human system, there has been comparatively but little advancement in respect to those of the brain. Doctor Stokes, in his valuable lectures on the Theory and Practice of Physic, has very correctly adduced the following circumstances as causes for such a state of things : "First, the great obscurity of the symptoms ; secondly, the want of correspondence between symptoms and known organic changes; and, thirdly, the necessarily imperfect nature of our classification of nervous diseases." Let us briefly examine these points. Now symptoms, according to this same author, "consist in certain changes produced in functions." But we have already seen that large numbers in the medical profession are wholly unacquainted with the real *functions* of the brain, and therefore they can not judge clearly and rationally of the *kind* or *degree* of functional derangement ; and hence the great obscurity attending the symptoms of diseases of this organ. We have, moreover, seen that *mental operations* constitute one of the most important functions of the brain — that the exercise of every individual faculty of the mind depends on a distinct cerebral organ ; but how little is definitely and practically known concerning the healthy or morbid manifestations of these faculties ! The knowledge that is already possessed on the subject is altogether too vague, indefinite, and speculative to be applied to any practical or useful purposes in medicine.

The fact is, the study of mental science, as based on the functions of the brain, must and will, in the process of time, constitute one of the most important features in the diagnosis and treatment of the diseases of this organ. As to the " want of correspondence between the symptoms and known organic changes," this is easily explained. It is more apparent than real ; for Nature never contradicts herself. It originates chiefly from a false view of classification of symptoms, and this, consequently, from an imperfect knowledge of functions. There may be, we admit, more than usual difficulty in ascertaining and settling this correspondence in the pathology of the brain, but a certain connection must necessarily exist between its functional derangement and change in physical structure, according to all the known laws which govern organic matter ; and we have not the least doubt but the precise kind and extent of this correspondence will yet be discovered and established.

The third difficulty in the way of understanding nervous diseases — viz., their imperfect classification — grows out of the two former, and can be rectified only in proportion as the functions of the brain become clearly and fully understood. The classification of no science whatever can be correct or perfect, unless it is based on a true interpretation of all the facts and phenomena in Nature appertaining to it. That the physician should be well acquainted with the most appropriate remedies in the practice of medicine, requires no argument to prove or enforce.

3. *The Importance of the Knowledge of Mental Science*

to the Physician. — This subject may be viewed under two general aspects : *first*, as connected with the duties which he owes to his profession ; and, *secondly*, in the relations which he sustains to the public. It will be seen from the preceding observations, that neither the anatomy, physiol- . ogy, nor pathology, of the brain can be fully understood without a knowledge of its functions, or, in other words, of mental philosophy ; moreover, that such knowledge is indispensably requisite, in order to understand correctly the diseases of the brain, as well as to perform successfully many operations in surgery for injuries of the head. This knowledge is especially important, inasmuch as the principal and almost the only means we have of ascertaining the affections of this organ is through the *kind* and *degree* of its functional derangement. We have no stethoscope to examine the state of the brain ; neither can we form or correct our diagnosis by the physical signs of auscultation and percussion ; neither is the brain, like most other parts of the body, susceptible of much pain from disease. Hence the great importance of understanding the functions of this organ, particularly of those portions connected with mental operations ; for the morbid or deranged manifestation of these will constitute the surest and most unequivocal symptoms of disease. To speak of mental excitement or depression in general terms, is not sufficient. We must know what *particular* faculty is involved, and *how much* it is affected. We might by such a course of diagnosis anticipate the very first symptoms of nervous disease, and thus employ remedial agents to much more advantage. It is not at all improbable but that a better knowledge of the functions and diseases of the brain will enable us to apply certain articles in the materia medica with far greater efficacy and success ; new medicines may in this way yet be discovered, or different combinations made of those already in use.

19

Again : A knowledge of mental philosophy can not fail
to be of great advantage to the physician in the treatment
of disease. That the state of the mind has a powerful
influence over the body, either for good or for ill, has been
universally acknowledged. It was remarked by Doctor
Rush, that " consumptions, fevers, convulsions, diseases of
the stomach and bowels, visceral obstructions, apoplexy,
palsy, madness, with a numerous and melancholy train of
mental diseases, are frequently brought on by the undue
action of the passions upon the body. ' All must admit,
that the faculties of the mind operate as powerful agents,
either as causes or remedies of disease. A multitude
of facts might be cited, where the exercise of certain
mental faculties has proved entirely effectual in preventing
or curing various affections. In this way, a salutary and
healing influence has been exerted upon the body when
other medicinal agents have been found utterly useless.
It is to this source that quackery and empiricism in medicine are chiefly indebted for success. It is by operating
upon the *feelings* of patients, that quacks perform so many
wonderful cures, and infuse such a magic charm into their
patent drugs. How important, then, that the regular-bred
physician should be thoroughly familiar with the nature
and application of an agency so efficient and powerful in
the treatment of disease! But it is not enough to be
acquainted with the powers or faculties of the mind, in a
vague, abstract, and general manner — such as love, hope,
joy, grief, fear, sorrow, anger, etc. etc. We must know
what *particular* organ in the brain is called into exercise
at the same time — what are the precise character and
strength of its mental faculty, and what are the most
appropriate motives to be addressed to it. We must
understand the nature and operation of those great laws
which everywhere invariably regulate mental manifestations, and be able also to explain every fact and phenom-

enon connected with individual minds. The physician, of all others, should be competent to do this to his patient, and a system of mental science based on the functions of the brain places within his power the means of obtaining such information. He would be able, in this way, to recognize at once the peculiar temperament or idiosyncrasy of every individual patient, and could thus take the advantage of a multitude of circumstances, of which he would otherwise be wholly ignorant. It is by pursuing such a course that a knowledge of mind can be rendered, in its applications, a "hundred fold greater," in the practice of the healing art, than the world has ever yet witnessed.

Again : The cultivation of mental philosophy is calculated to exert a beneficial influence upon the progress of medicine. Our present limits will permit us to notice only a few of the advantages to be derived from this source.

First, It will tend to do away with many groundless theories, hypotheses, and speculations, which, more than any thing else, have retarded the progress of this science. A large number of the works on medicine are comparatively worthless, because they are, in a great measure, filled with mere rubbish of theory, controversy, and the opinions of men who can not be considered as competent judges or safe guides. These theories are partly of ancient and partly of modern origin. The cultivation of medicine formerly partook very much of the manner and spirit in which metaphysics were studied — dealing in abstractions and generalities, without sufficient regard to facts, or the nature of the evidence upon which they were professedly based. The inductive philosophy, introduced by Lord Bacon, produced a great revolution in the study of medicine, and pointed out the true mode in which every department of this science should be cultivated, and by means of which most of its discoveries and improvements, for the last fifty years, have been effected. Now, a system of mental science, based on the function of the brain, is founded upon the most extensive induction of facts, and enforces at every step the absolute necessity of observation and experiment. It will, moreover, tend to bring into constant exercise the *observing faculties* of the medical student, and render him exceedingly cautious that his conclusions are always founded upon correct data. It will thus prevent too hasty generalization in medicine, and eventually become a standard to test the truth or falsehood of every new doctrine which claims to be based upon the laws of physical organization.

Secondly, Such a system of mental philosophy will enable us to test the real merits of the *opinions* of men, and decide how much weight should be given, in matters of science, to mere human authority. There are four classes of persons whose opinions in medicine should always be scrupulously examined, and on certain subjects they should be set aside, no matter how extensive their experience or profound their attainments; the difficulty arises from the *peculiar constitution* of their minds. The first class may be characterized as possessing very strong observing faculties, with deficient reflective intellect; these may observe, collect, and understand facts to any amount, but can never perceive or comprehend the force of *principles*, because they are *naturally* deficient in the powers of analysis and ratiocination. Wherever *general principles* are concerned, this class are not, therefore, competent judges. The second class of persons possess minds of a directly opposite character, having strong reflective faculties, but weak perceptive intellects; such individuals are not much given to observation themselves, neither can they appreciate the importance, or see the bearing, of *facts* in reasoning. They are inclined to dwell almost exclusively upon *general principles* and *abstract relations*, and not unfrequently become very speculative and theoretical in their views. Consequently, their opinions on all *practical* subjects must be received with much caution. The third class may be described as possessing, naturally, such an inordinate degree of self-conceit and tenacity of will as to render them blindly obstinate and wilfully set in their own way. They are always self-opinionated and unwilling to examine new subjects, or alter any views which have long been entertained ; and when their minds are once made up, no force of argument, or amount of evidence, will induce them to change or modify their opinions, simply because they *will* not be convinced. In the fourth class we would include those who are considerably advanced in life, and whose habits and modes of thinking have become so fixed and settled as to run almost necessarily in one circle or channel. Such are the nature and organization of the brain, on which the exercise of every mental faculty depends, that it is very hard, if not impossible, for elderly persons to canvass properly and rationally the merits of new discoveries. It is true, individuals of this class may occasionally keep up with the times and obtain a very good knowledge of all the passing events of the day ; but it is rarely — very — that an entire revolution or radical change takes place in their *opinions* on any important subject with which they have been constantly conversant for many years. This principle holds good both in relation to philosophy and religion, as well as the arts and sciences. Such is the testimony of all past history on this subject. We yield to none in our respect for age, as well as our confidence in the judgment of those of long and successful experience ; yet we do say, that the opinions of men passed the middle of life should have comparatively but little weight in settling the claims of *new* discoveries and improvements. We verily believe that not only medicine, but the progress of civilization, as well as of the arts and sciences generally, have been seriously retarded by giving an undue importance to the mere authority or opinions of such men.

Thirdly, The study of mental philosophy will eventually rectify or counteract the injurious effects of nosology on medicine. It has been a most unfortunate thing for this science, that its teachers should ever have laid so much stress upon the mere nomenclature and verbal description of diseases. In the first place, in order for such a course to be correct, it presupposes that the nature, causes, and symptoms of disease are already clearly and fully understood; and in the second place, that no change can be effected in these, either by time, climate, or other circumstances ; and, lastly, that all individuals will look at these facts through the same medium, and arrive at precisely the same results ; either of which conditions is absolutely impossible as well as absurd. Now, a nosological classification of disease, based on premises so false and erroneous, could not fail to have a most disastrous effect on medicine, and such has actually been the case. It has always operated as a serious barrier to any change or improvement ; it has filled volumes on medicine with words comparatively destitute of ideas ; it has cultivated the memory and fostered the credulity of the student at the expense of his judgment and independence, and led him as a physician to prescribe for the names, rather than the symptoms, of disease. Now, a system of mental science, whose invariable motto is, " *Res non verba quæso,*" will lead to a more correct use and interpretation of language. It will teach us that words are the mere exponents of ideas, and should never be employed without clearly expressing some idea or stating some fact. It will show the absurdity of attaching fixed names and stereotyped descriptions to phenomena, the features of which are constantly changing, and so blended·with each other that no distinct lines of demarcation can possibly be drawn between them. It will constrain the student to observe and think for himself, and not rely so much on the opinions of others ; it will compel him to study the great book of Nature, rather than the productions of men. The immortal Hunter used to exclaim to his class, while pointing at the human body, " I never read — this is the book that I study, and it is the work which you must study, if you ever wish to become eminent in your profession."

That a knowledge of mental science is important to the physician in his relations to the public, may be rendered obvious by numerous other considerations, aside from its bearing directly on his professional duties. We have already seen that such knowledge is not only salutary, but absolutely indispensable, in order to understand correctly many diseases to which the human body is subject ; moreover, that it is of the highest importance in the treatment of diseases that the physician should be thoroughly acquainted with the faculties of the mind, and the laws which regulate their development, as connected with the brain. Now, as the lives and the health of the community — objects the dearest and most sacred to every human being — are frequently entrusted to the care of the physician, not only the dictates of philanthropy, but the claims of justice, require that he should make himself fully acquainted with all the remedial helps and agents in his power, which are calculated either to restore health or prolong life. It is also a duty which he owes to his individual patients and the public generally, to employ his medical knowledge and exert his personal influence to prevent, as well as cure, disease. This should be one of the leading objects of every well-educated and liberal-minded member of the medical profession. But in order to do this successfully, the community, as a body, must be made far better acquainted with the laws of the animal economy, and the means of preserving health, than they now are. Formerly, it was supposed that man had but little control over the causes of pain, disease, and death ; some considered these afflictions as the mere results of chance or accident, while others viewed them as the visitations of a "mysterious Providence," and all apparently thought little, and practically cared less, about informing themselves on the subject. Now, it is found that disease and premature death are the penalties of violated laws — laws which it is the duty as well as the interest of all to study and obey.

There is no question but that disease in a multitude of instances might be prevented — that a vast amount of health might be saved, and the lives of many individuals be very much prolonged, by a more general diffusion among all classes of a knowledge of physiology and hygiene. But before mankind will ever pay that attention to the laws of the animal economy which their nature and importance actually demand, they must see and realize the entire dependence of all mental manifestations upon physical organization. The omission of this fact, whether it has been through ignorance or neglect, is one of the principal reasons why these laws have hitherto been so little appreciated or applied, both by the learned and the unlearned. Now, a system of mental science, based on the functions of the brain, is calculated more than any thing else to impress upon individuals, and the public generally, the importance of attending to those subjects which will vastly augment human happiness, by the prevention of disease and the promotion of health. And just in proportion as the principles of this science become understood, in the same proportion will individuals be induced to study the nature of their own constitutions, and yield obedience to the laws which govern them. For it will be found, by taking this view of the subject, that

all possess within their own power the means of self-
preservation and improvement, to a far greater extent
than has ever yet been understood in past ages, or is even
now conceived of, by the great mass of the public. When
we consider that all the manifestations of the mind depend
on the brain, it becomes an inquiry of the highest moment
to ascertain what are the causes or instruments operating
to affect its development, and what may be the degree of
influence which we can personally exert over these
agencies. It will then be made to appear how powerfully the character of every human being is affected by
physical organization — that the degree of his adaptation
to the enjoyment of the social and domestic relations, his
desire and capacity of elevation as a moral and religious
being, and also the amount of his intellectual ability,
depend in a great measure on the brain; then, and not till
then, will the attention of the public be suitably waked up
to the importance of this subject. And of all others, it is
the peculiar province, and may we not add the imperative
duty, of the physician to be foremost in imparting this
knowledge, and to take the lead in effecting a result so
desirable and philanthropic.

But these principles have a wider range, and embrace
far higher objects, than mere physical health or individual
enjoyment. They have an important bearing on every
thing which affects the interests of the human mind in
this world, as well as its preparation for the next. The
will of God, as revealed to man, may be found engraved
upon His works, as well as in His written word; and the
laws of the former are as binding and obligatory on His
creatures as the injunctions and requirements of the latter.
Before even Christianity can become practically what its
Divine Author intended, rather before its fruits will ever
be exhibited in the conduct of men in all that beauty, consistency, and perfection which characterized its great

Exemplar while on earth, the laws of the mind must first be correctly interpreted and obeyed.

It should be remembered that these principles, though they had their origin with the creation of man, have but recently been brought to light and made evident to the human intellect ; and although they are considered as fully proved and established as the facts of Chemistry or Geology, by all who have thoroughly and impartially examined them, yet the extent to which their truth is admitted, or that an application of them has actually been made, is very limited. This great work, therefore, remains yet to be done, and no small share of the labor belongs appropriately and necessarily to members of the medical profession. For the studies and pursuits of no other profession, or class of persons, are so nearly and intimately connected with mental science ; this fact must be obvious from the exposition which we have already given of its principles. But aside from the superior advantages which the physician enjoys of studying the physiology of the brain, and understanding the various conditions that influence or modify its functions, the peculiar duties of his profession place him in the most favorable circumstances possible for acquiring a knowledge of human nature. In the language of Doctor Spurzheim, " No one has such opportunities of observing men at all times, and in all situations. He alone is present during the night and the day, to witness the most intimate concerns and the most secret events of domestic life. Good and bad men, when sick, with difficulty conceal from him their true sentiments. To such a man, as knowing all that belongs to our nature, we unfold the most secret thoughts, and we acknowledge our frailties and our errors, in order that he may judge truly concerning our situation. There is, consequently, no man more called upon, no man more necessitated, to study mankind than the physician." Says

Doctor Rush, "It is the duty of physicians to assert their prerogative, and to rescue mental science from the usurpations of schoolmen and divines."

But it is when we consider the great variety and extent of the applications of this science, that its cultivation becomes so important, and urges its claims on our attention in a manner superior to all other sciences or subjects of human research. It would require volumes to unfold all its numerous and varied applications, only a few of which can here be mentioned. It points out the only true mode of education — physical, intellectual, and moral — that deserves the name. It has already shed a vast deal of light on the nature and treatment of insanity, thus bringing "joy and gladness" to multitudes whose situation for ages has been considered hopeless and irremediable. It is destined also greatly to reform and perfect our present systems of medical jurisprudence, criminal legislation, and political economy, as well as our social, civil, and religious institutions. It lays the only foundation for a system of ethics and morals — being the true exposition of the faculties and laws of the human mind. It is the "handmaid of religion" — the "elder revelation of God," and will eventually become "the *philosophy* which the world for centuries has had only in expectation."

The Normal Standard for Motherhood.*

BEFORE entering upon the discussion of this subject it seems proper that some general remarks be
made. The title of this paper is unique in its character
and requires explanation. The term normal implies
that there exists some rule or guide for reference. In
anatomy and physiology it signifies a sound structure and
healthy function, but in medicine it expresses a high state
of sanitation. The term standard is intended to designate a
certain type of physiology which furnishes us the best
possible organization for *motherhood*. We believe the
Creator has organized in the human body a fundamental
law for its increase ; and inasmuch as the organization of
woman has chiefly to do with this law, the discussion is
confined to her organism and agency.

Many illustrations of this law have been given, but the
most striking evidence may be found in applying certain
tests to the body itself. In examining various parts of
the system and descriptions of the same, we trust the
whole discussion, in fact and language, will be made intelligible to all. These tests have been applied with as
much skill and propriety as possible, trusting that the
evidence of this *"law of increase"* will become manifest
throughout the discussion.

* This article appeared in the *American Journal of Obstetrics and Diseases
of Women and Children*, April, 1876, William Wood & Co., N. Y. This
paper attracted much attention at the time, and was commended in high
terms by several of the most prominent medical men in the country.

In considering the foundation of this law, and the advantages to be derived from it, our remarks will be confined to a few points of view only, presenting a meager outline or brief synopsis of the subject. As the field of inquiry is comparatively new, and but little can be gleaned from medical works bearing directly on the subject, we would bespeak, in its discussion, the charitable consideration of the reader.

If the production of offspring is a primary design in the organization of woman, upon what particular type of development is the law found to operate best, or in its highest degree? That there is a difference, a wide difference, in the fertility of women, must be admitted ; a difference, physiologically,in the susceptibility to conception, in the effects of pregnancy, in the ease and safety of delivery, in the physical qualities for nursing, in the constitutional healthiness of offspring. In what, then, does this difference consist? Can it be confined wholly to the reproductive organs, or to the pelvic region alone? To settle the question we naturally seek some standard to which we may appeal ; and both Nature and analogy would lead us to the conclusion that such a standard or model certainly exists somewhere, and that we shall not seek for it in vain. Reasoning *a priori*, we should naturally infer that it would be found in the highest type or most perfect organization in structure and function ; for such are the nature, importance, and complication of forces required in propagation, that, for its successful results, it seems to demand the aid of every part of the system. This is certainly the first, the highest, and the most important law in the whole animal economy.

If we study the operations of Nature in the framing and government of organic bodies, we never find great principles or laws based upon any particular parts of the system or class of organs, neither upon inferior or imperfect structures. All the primary laws of Nature and the fundamental principles of science are exemplified in, and illustrated by, models of faultless forms and full development. The laws that govern the human system can not be an exception to this rule. If Nature has established such a law of propagation, it is of the highest importance that it be known and understood. While the recognition and knowledge of it would be fraught with the greatest possible interest and benefit to the community at large, it must prove of incalculable value to the medical profession. The law here proposed will be found, we believe, to rest, not upon mere theory or vague speculation, but upon positive facts ; and, if so, to lead, in the broadening of our knowledge and our researches, to results of a practical and valuable character.

While many facts and arguments may be deduced from the general principles of physiology in favor of such a law, there are four distinct points of view, around which they may properly be gathered, and, in this way, be brought out and illustrated in a clearer and more forcible manner. These points are: The pregnant state ; the mechanism of labor ; the qualifications of a nurse, and the character of offspring. If there is a general law of propagation, a normal standard in the organization of woman, based upon the principles of physiology, it will certainly aid us to a better understanding and knowledge of those important changes through which she must pass in child-bearing. Let us then briefly review the leading facts or phenomena in each of these changes, and see what lessons they teach.

One of the most eventful and trying changes the human system can possibly pass through is that of

PREGNANCY.

This state causes many physical changes — some simple and safe, others complicated and occasionally dangerous.

The primary changes can not properly be considered actual disease, but rather functional derangements. In works treating of diseases of women we generally find a long chapter, headed "Diseases of Pregnancy," discussing from forty to fifty distinct complaints arising from this source. But pregnancy in itself can not be considered strictly a morbid or diseased state, inasmuch as propagation, in its normal effects, must harmonize with the principles of physiology. Montgomery, one of the most distinguished writers on this subject, makes this significant remark : " If, with a few, pregnancy has deserved the name of a nine months' malady, fully an equal number suffer little or no inconvenience, and with some it is a period of decided improvement in health ; moreover it appears, from all experiences, that women who bear children generally enjoy more even health, and are less disposed to disease, than those who lead a life of celibacy, or who, having married, remain unfruitful." Now, why should there be this difference ? Why should some women suffer so much from the pregnant state ; others so little, and others still improve by it in their health permanently ? It may be said this depends upon differences in constitution — the pregnant state, in one sense, agrees with the constitution of some women, but disagrees with that of others. What, then, is that agreement ; what is the type or character of those constitutions with which the pregnant state harmonizes ? Is there not some law or standard by which these can be tested or explained ? In the very nature of things there must be, in these matters, the observance or violation of law. Such changes can not come from chance.

Every experienced physician knows full well that there is a great difference in women as to the effects of pregnancy, and that these effects are various and occasionally very marked. Sometimes the change may affect this organ

— sometimes that ; and again, almost every organ in the
system becomes more or less affected.   In some cases the
very first stage of this change operates unfavorably ; it
may induce a little nausea or slight headache, or it may result in the most violent inflammation or convulsions.
While some women may be benefited in their health
from the change, and their constitutions actually improved
by child-bearing, with others it is the commencement of
suffering and disease, resulting in impaired health and
not unfrequently a broken-down constitution.   Now why
should there be these differences, why these disturbances?
What are the causes, the constitutional weaknesses, the
particular predispositions ?   If propagation is physiologically a normal function of woman, why these pathological
changes ?   What laws have been violated ?   And why
should there be such a marked relation or sympathy between this change in the reproductive organs and other
parts of the body ?   The very fact that one organization
is found more favorable for child-bearing than another,
implies that there is somewhere a normal standard ; and,
if so, let this change or improvement be carried to a
standard of organization where the least bad effects possible arise from the pregnant state.

Let us make an application of this principle to different
types or kinds of organization.   In tracing out the effects
of the pregnant state, we find some difficulty, from the
fact that by this change in the uterus it works in three
ways — by attraction, by sympathy, and gradually by
mechanical pressure.   One of its effects is to change the
circulation and the direction of the nutritive force.   Thus,
where in certain parts of the system there has been over-
action or excessive excitability, perhaps a strong predisposition to, if not the actual existence of, disease, the
pregnant state, in changing the circulation by withdrawing
from these organs a certain amount of blood and nutrition,

actually improves the health, and in some instances undoubtedly prolongs life. Here an attempt is made by a
natural law to correct weaknesses and restore health, or in
other words, to bring about a more even balance or better
harmony of action in the whole organization.

There are other cases where the weaknesses or excesses
are so great, or the disease has been carried so far, that
pregnancy makes the attempt to change this state of organization and not only fails in so doing, but perhaps indirectly aggravates the difficulty. In all those cases, however, where the health of women is improved by pregnancy, it is accomplished, we believe, by so changing the
current of the vital forces of the system as to bring about
a more equal circulation, a better balance of organization,
as well as harmony of function throughout the whole body.
What, then, is the inference, or what lesson does this class
of facts teach ? Is it not clearly this : that the better and
more evenly balanced the structure of the whole body is,
and the more perfect the action of its machinery, the less
disturbance will be produced by pregnancy, and the less
harm or inconvenience result from it ?

Let us look at different types of organization. The more
nervous and sensitive a person is, the greater and more
marked is the effect of pregnancy. In such cases, generally, the change is sooner discovered, and the signs or indications arising from it are more decided and positive.
In some cases, where there is a great preponderance of
the nervous temperament, pregnancy, by changing and
equalizing the action of the nervous system, may improve
the general health and constitution. In other cases it may
increase and intensify the nervous activity or excitability,
and thus affect, more or less, the disposition and temper
of the individual. Now and then a case occurs where
pregnancy has a singular effect upon a nervous temperament, to disturb and excite the patient, and sometimes

20

even cause mental derangement. In all such cases, if the exact physiology and pathology of the brain and nervous system could be ascertained, we should find some peculiar sympathy in the relations of the nervous system, or some singular idiosyncrasy of organization, in these persons. But such cases do not often occur and are exceptions to the general rule. Such changes of conduct or exhibitions of character do not occur without a cause; and when the cause can be ascertained and satisfactorily explained, instead of conflicting with, or furnishing evidence against, general laws or principles, the history and explanation will rather serve to confirm and strengthen the laws.

There are many slight disturbances occasioned by pregnancy in the action of the stomach, bowels, heart, lungs, and nervous system, which, so far as they prove any thing, show a well-balanced organization in those cases, and also that no marked weaknesses, defects or disease existed. But occasionally the stomach is greatly disturbed, which leads to serious and dangerous disease. A careful examination into such cases will show, we believe, a remarkable sympathy, or sensitiveness, between the state of the stomach and the action of the uterus or other organs. It may show, too, that the individual had suffered more or less, for a long time previous, from dyspepsia or indigestion.

Sometimes great physical changes are occasioned by the pregnant state: the woman, occasionally losing flesh and strength, continues to waste away till she can scarcely go through the regular period of gestation, the nutrition going mostly to the child, and the whole change being caused by some defective or unnatural action of the digestive organs; but, more often, the woman gains flesh and strength, becoming plethoric, and, as it may be said, corpulent. In such cases, the stomach and digestive organs act too vigorously,— manufacture too much nutrition and blood, certainly for the mother, though perhaps at the expense. somewhat of the child.   This change is decidedly unfavorable, resulting not unfrequently in convulsions or violent inflammation.   The causes of such a change are not easy always to understand, but indicate that there must have been some radical defect in the organization, or something wrong in the habits of the individual.

Sometimes the liver and kidneys are so affected by pregnancy as to change the quality of the blood, resulting in anasarca, thereby enhancing the danger of the condition, and sometimes resulting in loss of life.   In some cases, it is thought that pregnancy, by sympathy or by some singular influence upon those organs connected with the process of digestion, produces albuminaria, causing most dangerous convulsions, and, in some instances, resulting fatally.   This disease, whether caused wholly by the pregnant state or not, is one of the most obscure and dangerous of all diseases.   It is not easy to trace out its exact relation to pregnancy, or to describe just what pathological changes had taken place in its preliminary stage.

Future researches in pathology will undoubtedly explain these causes and changes — and, we are confident, they will also show that there were some conditions in the organization, or in the habits and health of the individual, existing prior to pregnancy, predisposing to this disease, so that this state of the system operated only as an exciting cause.   As yet pathological inquiries have not been carried far enough in this direction ; but, when thoroughly prosecuted, we believe that they will show that the sad results of this morbific state or diathesis may be in a great measure obviated, and perhaps show that it is not chargeable to pregnancy alone.

There is another class of complaints, arising from pregnancy, caused by mechanical pressure, interfering with the circulation, especially in the lower extremities ; and sometimes this pressure operates unfavorably upon the natural action of the bowels and stomach, as well as upon the functions of the liver, heart, and lungs. This result of the pregnant state can not well be obviated, or much relieved by any mechanical treatment, as it arises from a want of proper development of the whole body, or from the too close relations of the internal organs, one to another. The disturbances from this source are more numerous, and their results more serious, we believe, than are generally supposed.

Again : Is there not a wide difference in the effects of pregnancy as found in different classes, nations, and races ? Are they not, as a whole, more marked and serious among the higher classes of society than the lower; in cities, than rural districts ; and less striking and troublesome still among women living even in a semi-civilized and barbarous state ? In fact, wherever the female organization is the most perfectly developed in all parts, and the functions of every organ are performed in accordance with its own inherent laws, are not the diseases of pregnancy the least marked and serious?

Now in all these changes and diseases, a careful investigation will show that, in case there were always a well-balanced organization and a healthy performance of the functions of the internal organs, we should have few diseases arising from the pregnant state. And all these complaints are found to diminish in number and severity, just in proportion as we find organizations approximating to more perfect standards. If, therefore, propagation is the normal state of woman ; and the more perfect her organization is, anatomically and physiologically, the less are the disturbances or diseases of pregnancy,— it certainly points to the fact, or affords evidence, that there exists in Nature what may be denominated a general law of propagation, based upon such standards of the system.

For a proper understanding of this law, it is highly important to bear constantly in mind, not only the striking differences in female organization, but to notice particularly the great changes it has undergone in different races and at successive periods of time. This fact will appear more obvious in the consideration of the next step or process in the development of the law, viz:

LABOR.

That there are wide differences among women, in the ease and safety with which they go through this process, all will admit. Now if this process of labor or delivery is natural to woman, is normal physiologically, why is it attended frequently with so much pain and difficulty, and with danger to life? In no other department of the animal economy, where the laws of Nature, in a normal state, are observed, do we find such pain, distress, and suffering. Do not these symptoms, therefore, indicate that the laws of organic beings, or the designs of Nature, have been in some way violated or perverted? It is true, some women go through the process without much suffering or loss of strength, while to others are allotted nights and days of pain, anguish, and distress; and it would seem that the latter class constitute, at present, the exception to the general rule. Teachers and writers on this subject have taken great pains to ascertain and describe what were the causes of so much difficulty and suffering in confinement, and to inquire what human means or resources of art could be employed to remove these difficulties and assist Nature in this work. To this end the anatomy and physiology of the pelvis have been carefully studied : the relations each part sustains to this process, — what were the precise functions of the uterus, — what should be the presentation of the child, — what obstructions, points of resistance, etc., existed.

No part of the body, probably, has been more carefully studied than that of the female pelvis, and no organs in the whole system perform such important functions as those located in this region. For better understanding and treatment, parturition has been divided into different stages, and its phenomena classified — such as natural and protracted, tedious and laborious, difficult and complicated labors, etc. etc. Special attention has been given to difficulties attending labor, such as position or wrong presentation of the child, the disproportion between its head and the pelvis of its mother, the imperfect and irregular action of the uterus, the rigidity of the os uteri and the soft parts, the necessity of using instruments, the danger from exhaustion, convulsions, hemorrhage, etc. These are the points or sources of pain, distress, suffering, and danger. In these eccentric or extreme organizations the greater are these marked peculiarities of the system, and where we can encounter a large part of the difficulties in obstetrics ; and the wider these divergences go, in any one direction, or the more marked these peculiarities are, the greater these difficulties. On the other hand, the nearer we approach a sound, well-balanced organization in all its parts, the greater the ease and safety in delivery. Every physician occasionally finds patients that go through this process with comparatively little trouble or difficulty. We find also among women all manner of differences in the process of labor ; and these depend mainly upon the kind or type of organization, together with the habits and health of the individual. Now, why these differences, why these peculiarities? Are they not deviations, more or less, from a perfect standard of organization; or in other words, are they not to a great extent abnormal? Are they not the effects or penalties of a law violated, or the result of an artifical — in some respects, an unnatural — life ?

If a test or direct application of the principles of physiology be here made, it may throw some light upon the subject. Should any class of organs, or some one temperament, such as the nervous, sanguine, or lymphatic, greatly predominate, its effects as a whole will in parturition be found unfavorable. If there is an undue predominance of the nervous system, there will exist far greater sensitiveness or susceptibility to pain ; and the process of labor may produce such a shock upon the brain and the nerves as to render recovery doubtful, if not impossible ; if there is an excess of the sanguine temperament, there will probably ensue a greater strain upon the action of the heart, and sudden change in the circulation of the blood, with increased danger of hemorrhage and inflammation. If the lymphatic temperament abounds, there is a sluggish state of the system, a lack of force and regularity in the contractions of the uterus, such a deficiency in general vitality and strength as to render parturition tedious, if not sometimes dangerous, from exhaustion. If the muscular tissue greatly predominates in the system, then we find, with violent pains, powerful resistance and rigidity of all the soft parts. It may be, these defects or peculiarities of organization will not show themselves so much in pain and difficulty of delivery, but their effects may become more manifest upon the system afterwards, or upon the character of the offspring.

There is a physiological condition or principle involved in labor, that is not, we believe, properly considered. We refer to a union or relation of forces in Nature, so that all parts of the system should act in harmony with each other, and in one single direction, when the object to be accomplished requires it. This principle in the study and practice of obstetrics has been, if we are not mistaken,

very much overlooked. Such are the nature and object of
propagation in importance and magnitude, that we should
expect aid from every part of the system, from every
tissue, nerve, tendon, and muscle. Parturition is certainly
one of the most important and complicated processes
in the fulfilment of the law. Now, while certain
organs are called on to perform their natural functions,
there should be no conflict or resistance from the action of
any other part or class of organs. But in an imperfectly
developed and unevenly balanced body, with a want of
harmony in the action of all parts, it is difficult, if not impossible, to obtain a union or conjunction of all the forces
of Nature in the most favorable manner. If the organization of woman, as found, is a deviation from the normal,
perfect standard, it could not be expected that all the
forces of Nature or the whole organism would aid in the
process of labor so favorably, or to the same extent, as
they would in a perfectly healthy or normal state. Hence,
in considering the causes of pain, the difficulties attending delivery, the force and relations of the whole system
should be taken into account, and our attention should not
be confined wholly to the pelvic region. It should be
borne in mind, too, that we are dealing with imperfect
organizations, where general law can not be fully applied.

There is another class of facts that have an important
bearing upon the subject. It has been remarked that
there are not only wide differences among women as to
pain and difficulty in parturition, but there are some
women, in every community, who suffer comparatively
little at child-birth. Now, a careful examination into the
structure and functions of the whole organism of such
women, we venture to assert, will show few excesses or
defects, but on the contrary, unusually well-balanced,
sound, and healthy conditions in every part and organ.

Now let this same principle be borne in mind, as applied to different classes, races, localities, and states of society. It may be difficult to collect here facts upon so large a scale, or to institute such comparisons, as would settle any general laws or principles ; but still information may, in this way, be gleaned, that will throw much light on the subject. One general fact is very obvious : from medical writers and travellers we learn that women, living in what is termed a state of Nature, suffer comparatively but little pain or trouble in parturition ; whereas all history testifies that this pain and suffering increases just about in proportion as civilization advances. Thus, in what may be considered a high state of civilization and refinement, not only more pain and distress are attendant on parturition, but increased difficulty and danger.

Among the North-American Indians, the inhabitants of Greenland, of Labrador, of the South Sea Islands, and among various classes in South America, of the numerous tribes of Africa and South-eastern Asia, child-bearing, we are informed, is accompanied at the present day with but little suffering or difficulty. There are undoubtedly individual cases in all these countries attended with distress and danger, but then these are the exceptions. In this general statement we do not deem it necessary to go into details of evidence by giving facts, making quotations from different writers, or furnishing various kinds of evidence. Many writers on obstetrics admit the correctness of these statements ; in fact, they are nowhere called in question.* Now why should there exist these distinctions

* It is more than probable that pain and difficulty in parturition are artificial, and are the consequences of civilization and refinement. For the human constitution, when not under the influence of these causes, will, *cæteris paribus*, be found capable of meeting and overcoming without any difficulty the ordinary changes produced by gestation and delivery. Of this abundant proof might be given ; for the female savage, wherever found, whether under the scorching heat of an African sun or beneath the rigorous sky of the unfriendly Labrador, brings forth her young without the assistance,

or differences in pain, suffering, and danger attending a process that is considered a natural, normal condition or function of physiology? In a primitive state of society, among people living in a plain, simple manner, with habits rude and uncultivated, we find but little distress or trouble attending propagation; but in society advanced in civilization, refinement, and culture, we find much difficulty, and not unfrequently danger, attending the fulfilment of this law; and the higher the degree, or the more advanced the state of this civilization, the more painful and hazardous are the chances. The question returns upon us, Why this difference? What are its causes? Are they necessary? Can they be explained? Can any thing be done to modify or to remove them? The inquiry naturally arises, What is the physiology of women living in the countries referred to, where the law of propagation is so easily complied with? May there not be found among them a better-developed physical systems, more evenly balanced in all its parts or organs, a greater harmony in the performance of all their functions, especially in reference to what may be termed the *primary laws* of Nature? Writers admit that there may be found, at the same time, individual cases of women living in these countries subjected to great suffering and difficulty in parturition, and sometimes danger in the process or from its effects.

of an accoucheur or midwife; but the reverse of this almost universally obtains among the females of the civilized world. These differences are most probably occasioned by the changes produced on the human constitutnio by civilization and refinement.

The mischiefs derived from the sources just mentioned are found to consist in the disposition to, or existence of, diseases, either general or local or both, in those which may affect the system in general, or those which may be confined to the uterus or pelvis in particular; in the introduction and continuance of certain pernicious customs, habits or modes of life, thereby inducing a preternatural degree of irritability, sensibility, laxity, or rigidity—and hence the physical necessity of pain and difficulty in parturition among the greater part of women in a state of civilization and refinement. — *Dewees' Essays, p. 25.*

The questions might arise, What was the organization of man at his creation? What were the designs, provisions, conditions, etc., with reference to his continuance? Whether we adopt the Scripture account of his creation, or the Darwinian theory, so-called, of his origin, what evidence can there be found that will explain or throw light upon any such general law of propagation? When man was created, according to the Scripture account, there is reason to believe that it was with a perfect anatomical and physiological structure in all its parts or organs, and that there was a perfect harmony in the performance of all their functions. And when the command was enjoined upon the original pair "to be fruitful, to multiply, and replenish the earth," the fulfilment of this command, with a perfect organization on the part of the woman, it is presumed, was not attended with much pain or difficulty.

But afterwards, in consequence of the disobedience of our first parents, the sacred Scriptures relate that the Almighty said to the woman, "in sorrow shalt thou bring forth thy children." The term sorrow, as here used, has received various interpretations. Some writers maintain that it refers exclusively to the mind — to mental acts — such as anxiety, fear, suspense, distress, etc., while others maintain that it implies also physical pain and suffering.

Then, again, the whole transaction is regarded by some as a judgment or curse pronounced upon woman for disobedience, which was to become universal, and continue through all time, without much relief or change ; while by others this declaration of the Almighty is interpreted as somewhat conditional in its application, — that sorrow and pain would follow child-bearing, because the laws of the physical system were violated, and that the amount of this sorrow and suffering would depend upon the manner and extent to which these laws had been violated. This view of the Scripture narrative is the most natural interpretation. It harmonizes not only with the character of God and our own moral sense of justice, but is confirmed by all the facts of history, as well as the principles of physiology. It implies distinctly that some changes would take place in the operation of this law, which would bring sorrow and suffering to woman. It is inferred that there was none or but little trouble of this kind in her primeval state. This change in the law resulted not from an arbitrary or vindictive spirit on the part of the Creator, but depended wholly upon the violation of physical laws by human agency; that just in proportion as man violated the laws of his own being, in the same proportion would there be sorrow attending his birth. Thus, in the various changes and deviations from this perfect physiological standard, to which the human body in all ages has been subjected, do we find an endless variety of sorrow, suffering, and hardship accompanying child-birth.

There is another point of view from whence important evidence may be gleaned. It is well known that there is a great difference in women as to the amount of prostration produced by the effects of labor, as well as in the length of time and manner of recovery. This depends much on the strength of the constitution, and also on the character of the labor. With some women, the shock is so great, and the exhaustion so excessive, that it requires weeks and sometimes months to recover, and occasionally there are some who never regain their former strength and health. There are others who go through the process of pregnancy and labor without much exhaustion, or even fatigue, and it is with great difficulty that they can be confined, after delivery, a week or ten days in bed. And they will go through this process ten, twelve, or fifteen times without apparently any injury to health or constitution — in fact, with scarce any loss of time, and not infrequently, after having a large family, they maintain remarkable health and live to a great age. Now, why are there such differences, such exhaustions, such slow recoveries, and, sometimes, permanent injuries of constitution ? Why do some women rally so easily and so soon after confinement, and seemingly improve, or at least hold their way, by every repetition of the process ? From a careful examination into a large number of such cases, we have always found that such women possessed a remarkably well-balanced organization, — not merely good health, devoid apparently of any particular weakness or disease, but a sound body, fully developed in all its parts and organs.

The differences in size and form of the female pelvis in different nations, and the changes in the form and character of this structure, in the same race through successive generations, from a rude to a highly civilized state, are very important considerations. It is maintained that the fœtal head also differs in form and shape ; that among a people highly educated the anterior lobes of the brain are larger, and that such change gradually takes place just in proportion to the advance of civilization. Von Franque, who has perhaps devoted more attention to this subject than any other writer, in accounting for the quick and easy labor in uncivilized nations, says : "We must not forget, in this question, the influence of culture, which certainly can not be estimated too highly ; so that, with increase of culture and super-refinement of customs, not only the most various diseases appear more numerously, but that also, in the same measure, the labors become more difficult and of longer duration ; that, especially, complications step in, which are conditioned by anomalies of the bony pelvis, and which are in general met with but rarely, almost not at all, in uncivilized nations."

While we admit that the changes in the shape and

diameters of the pelvis effected by culture, refinement, habits, fashions, etc., of civilization do greatly increase the difficulties of parturition, may not the changes in other tissues, or parts of the body, from the same causes, increase also these difficulties? The muscular power of the uterus is certainly not dependent upon the size and shape of the pelvis, neither is the strength or power of endurance of the whole body. As the quotation from Von Franque states, "various diseases and other complications" — and may we not add weaknesses too?— have been introduced by these causes, which greatly increase the pain, difficulty, and danger of parturition. In fact, if all these difficulties, including the suffering, exhaustion, hemorrhage, convulsion, puerperal disease, etc., were carefully analyzed, what proportion of these originate solely from the bony structure? While no distinct line can be drawn between a portion of these and their primary cause, yet if a survey of the whole could be correctly made, and their causes defined, we question whether one-half of them would be found to arise exclusively from the pelvic bones. But it is in the matter of conception, pregnancy, gestation, lactation, etc., that these changes produce their greatest effects on the physical system.

It should be borne in mind that the changes here referred to do not grow out of a true, healthy civilization, but from an artificial type, from wrong habits, pernicious customs and fashions, from an unnatural culture and refinement, where the laws of health and life are altogether too much violated. It should also be borne in mind that these changes have not been the growth of one generation, but of many ; and thus, by the laws of inheritance, they have become greatly increased, and their effect intensified.

Without going further into details on these points, let us sum up what seem to be the general facts upon the subject. It is admitted that there are wide differences among women as to the pain and difficulty in parturition. It is found that in the ruder portions of society, and among the semi-civilized and semi-barbarous nations, very little pain or trouble, comparatively, is experienced in child-bearing. From the Scripture narrative we have good reason to believe the organization of woman at creation was such that she suffered little pain from this source ; but afterwards a change occurred whereby her liability to pain and suffering was greatly increased. All history shows that, in proportion as the human body has changed by artificial habits and vicious practices, woman has been subjected to greater and severer pain, as well as difficulty, in child-birth. Facts also show that the further artificial habits, luxuries, and fashions are carried, the greater the distress, difficulty, and danger in child-birth. Now what lessons do these facts teach ? Do they not plainly indicate that there exists somewhere a normal standard, established by physiology for propagation ? Do they not teach that the nearer the physical system of woman approaches that standard, the less pain and suffering she endures ? If there is, then, such a standard, what is it — in what does it consist ? We answer, a well-balanced organization, sound in structure and harmonious in function, in which every tissue and organ are developed to the highest extent that is compatible with the healthy performance of their functions.

The next stage in the observance of this law is the dependence of the infant for nutrition upon the mother, or in other words,

THE QUALIFICATIONS OF A GOOD NURSE.

There must be in this respect, between the two, natural adaptation or harmony of relation. According to the laws of Nature, when properly observed, we find that wherever

she makes a demand, she is also sure to furnish a supply.
Her laws, too, when correctly interpreted, are found not
only to harmonize with each other, but are always complete in design — never disjointed or fragmentary. Thus,
lactation, in the natural order of things, must follow
parturition, as much as that process must necessarily
follow the pregnant state. It was evidently intended by
the laws of Nature that the child, for months at least,
should be supplied with nutrition wholly from this source.
No fact in vital statistics is more firmly established than
that, in order to save life and promote health, the infant
should be nursed at its mother's breast. The ingenuity
of nurses and physicians has been taxed to the utmost,
the principles of chemistry and the results of experiments
have been brought into frequent requisition, but no substitute can be provided equal to pure breast milk. Nature,
in her normal state or highest development, we believe,
has made ample provision, in the organization of woman,
for nursing her offspring. But in order to provide this
nourishment pure in quality and abundant in quantity, she
must have a well-balanced organization, especially a good
development of the lymphatic and sanguine temperaments,
together with vigorous and healthy digestive organs. The
mammary and other glands should be neither too large
nor too small; the powers of mastication, digestion, and
assimilation must not be deficient, must be equal to the
demands which Nature makes upon them in this direction.
If there is a great predominance of the brain and nervous
system, and a constant strain is made upon those parts,
thus requiring a large amount of nutrition and exhausting
the vitality of the system, there must be a failure in lactation. On the other hand, if the organization of woman
partakes too much of the lower animal nature — abounds
in flesh — if she is physically large and unduly corpulent,
the powers of lactation here fail, the organs of diges-

tion and assimilation may work vigorously, but the nutrition will go to the mother and not to the child. A careful
examination into the physical qualities of women who
nurse their offspring best, will show a natural fitness or
adaptation for this work. This same law holds good in
the animal creation. There, from pecuniary considerations, it has been made a special study. Experiments
have been tried without number, and observations made
upon the largest scale ; no pains or expense has been
spared in devising ways and means whereby the best and
largest quantity of milk could be obtained from domestic
animals for the use of man. But how little interest or
attention has been devoted to the subject of obtaining a
proper supply of human milk for infantile life! Is not
the life of the infant as valuable as that of the adult?

As to this matter of nursing, a variety of opinions have
been entertained by different writers. It has long been
observed that there were great differences among women
as to their qualifications for nursing; some furnish an
abundance of milk, some only a partial supply, while
others can furnish but little. Instead of studying into
the physiology of women, and inquiring what there
was in their organization that made these differences,
attention has been devoted almost exclusively to the
means of providing an artificial supply. Upon examination into the instructions and directions on this subject,
as found in books and lectures, there seems to be something wanting ; the obvious principles or teachings of
physiology have not been properly expounded in their
application to this function ; neither has it seemed to be
considered that the laws which govern, in this respect,
the animal creation, are precisely the same as those that
govern the human race. In confirmation of our statement we will make a quotation from an address before a
large body of physicians, by a professor of obstetrics and

21

diseases of women in one of the oldest and largest medical schools in the country. Says this professor: " Why do American-born females make such poor wet-nurses compared with the immigrant from Ireland or Germany? After nearly thirty years of practice I can not answer the question. That it is the fact, few practitioners in our large towns and cities doubt. Allow that some women with us, as with foreigners, object to being bound to their children's cries, yet the mass of American females are totally unable to act the wet-nurse with success." It is not three years since this statement was made and published. This is, we imagine, a more candid comparison than many medical teachers or writers would care to make. But it is the truthfulness of the statement, and the explanation offered, to which special attention is here called. Why should there be, in this respect, such a difference between American women and the Irish or German immigrant? Why should New-England women of the present day differ from their mothers and grandmothers, who found but little difficulty in nursing their offspring? Formerly it was a rare thing in New England for a mother to be obliged to resort to a wet-nurse or to feeding by hand. But now it is certainly within bounds to state that not half the New-England women in cities and large towns can properly nurse their offspring. It has been supposed, however, by some that all our American women can nurse their offspring just as well as not — that the disposition only was wanting. But this is found practically a great mistake. While there may be cases, here and there, of this indisposition to nurse, it is a fact that large numbers who are anxious to nurse make the attempt, but fail. They find, after repeated attempts, that their milk does not satisfy the child, or that it does not thrive; that there must be deficiency in the quantity or defects in the quality of the nourishment. In many cases, after trying the experiment for weeks or months, they are compelled to give up nursing entirely, while others, depending partly upon nursing, resort also to artificial means for feeding the child.   So impressed have writers been on this subject, and also practitioners of medicine, that the nursing of offspring harmonizes with the laws of physiology, and, as a general rule, proves beneficial to the health of the mother, that they uniformly advise that the mother should by all means nurse her child.   This has always been a favorite theory with obstetricians, and its correctness has been confirmed by the results of experience and observation gathered from all quarters.   Such we should expect from the obvious teachings of physiology, and it certainly accords with the common judgment of professional nurses and mothers themselves.   But if the principle here laid down is correct, why should nursing be so often attended with pain and difficulty? That there is sometimes a defect in the form of · the nipple, and the act of nursing becomes very painful, we easily understand.   There are some cases where the act of nursing causes the most painful sensations, extending through the breast to the spine, and from thence through almost every part of the body.   There are cases, too, where, after a most faithful trial, nursing actually disagrees with a woman and proves, in a variety of ways, unfavorable to her health, so much so that she is compelled to give it up for the preservation of her own life. There are other women, at the same time, with whom it agrees—is found to improve the health through the whole process—that they were never so well as when nursing, even though this process should be repeated from the tenth to the fifteenth time.   Now, why do we find such a difference in the effects of nursing? Why should it ever be attended with pain and difficulty?   Why should it injure the health of one and improve that of another? There must be causes or reasons for these.

Now, these facts as to the inability for nursing may be found, perhaps, more common in New England, though cases of this kind are not wanting in other portions of our country, both among the immigrant as well as native-born women. Such incapacity has been found to exist, more or less, in all countries and among all classes and all races. It has not been confined to any age, climate, or country, or to any tribe, race, or color. But formerly this inability was not so common — occurred only occasionally, and when partial, did not attract any attention. As long as such cases constituted exceptions to the general practice, they did not create much interest, or lead to remarks or observations on the subject. The same fact is true at the present day in respect to the German, English, and Irish ; a large majority of these women nurse their offspring — those who can not, or do not, constitute the exception. But in New England a gradual change has been taking place : the fact has become more and more apparent, that large numbers of women can not nurse their children, — so much so, that in certain localities or classes those who do are beginning to constitute the exception.

The questions may very pertinently be asked, Why this change? Why this anomalous state of things? Why do we find so many exceptions in the observance of one of the most important functions of the system? What is there here abnormal and unnatural? These inquiries open up the whole subject as to what constitutes the physical qualities of a good nurse — not merely in New England, but in all countries, and among all classes and people.

It is evident that the whole matter of the mother's affording proper nutriment to her offspring at birth, and afterwards as long as its nature requires, is governed by some fixed laws. The last is indisputable, and there can be no question or haphazard about it. In the very nature of things these laws must have their foundation and support in physiology. As in other organic functions, so in the secretion of milk, there should exist the requisite organs in good development, and these should have their proper share of aliment and support.

The organs classified particularly under the lymphatic and sanguine temperaments must be not only well developed, but other parts or organs of the system must not be found altogether disproportionate to these. In this case those portions of the body that are predominant require an undue share of nourishment; if it should happen to be the nervous system, and particularly the brain — as this tissue requires relatively a much larger proportion of nutriment than any other — such an organization would be poorly fitted to afford proper aliment for offspring. The more carefully all the physiological developments or conditions requisite for a good nurse are investigated, the more convinced we shall be that they depend not merely upon what may be considered a sound and healthy body, but upon one well-balanced, evenly developed in all its parts. As far, then, as lactation is concerned, this type must be considered its normal standard.

That the human body has undergone changes from time to time, all will readily admit. Many of these changes, occasioned by the artificial habits of life as well as by the fashions of the day, are found not only unfavorable to female health, but must prove decidedly injurious to the race. Nearly forty years ago Sir Astley Cooper made this statement : " It is melancholy to reflect that a life of high civilization and refinement renders the female less able to bear the shock of parturition ; it has a tendency to lessen her attention to her offspring and really diminishes her power of affording it nourishment, so that she is often a worse mother in these respects than the female of the

middle ranks of life, or even the meanest cottager." This
remark was undoubtedly made as the result of extended
observation and long experience, many years ago ; and it
implies not merely a change of disposition, but also a
change in organization, from the fact that such mothers
could not properly nurse their offspring.   Sir Astley
Cooper observes that the proper development of the
mammary glands is often prevented by a constant pressure. We might go further, and say that continued compression of the chest and abdomen is calculated to impair
the development and healthy action of the lungs, the
heart, and digestive organs, as well as those in the pelvis.

If we consider that this compression commences with
the girl or young woman, when the system is in a state of
growth and most susceptible of change — that it may be
continued for a series of years, and, by the laws of inheritance, intensified, it shows very clearly how such
effects upon the system disqualify women for some of the
most important duties of maternity.   A great variety of
causes, other than those here stated, might be adduced
to account for physical changes of constitution, or
changes which might especially interfere with the lacteal
functions.   Among these causes may be mentioned
educational pressure, constant excitement, depression of
spirits, too much society, hard work, great exhaustion, etc.

In the matter of nursing, much depends upon the daily
habits of the individual, the kinds and quantity of food
consumed, the nature of drinks taken, etc.   While these
agencies have, for the time being, a marked influence
upon lactation, it is the particular type or standard of
organization most favorable to nursing, that constitutes
the present object of our inquiry.   We have stated that,
in the matter of nursing, there is a great difference
between the women of New England at the present day
and the early settlers. That there has been here a decided

change in female organization within fifty or a hundred years, there can be no question. Formerly, there were more muscle, a larger frame, greater fulness of form, and a better development of all those organs that are classed under the sanguine and lymphatic temperaments. The brain and nervous system relatively were not especially predominant ; neither were they taxed continuously or excessively above any other class of organs. Those of the Germans, English, and Irish who best nurse their offspring at the present day, possess an organization similar to the one here described. If an inquiry could be thoroughly prosecuted in any tribe, race, or people, and the individuals or classes that were found most successful in nursing their offspring could be picked out, we should find that they possessed an organization much alike, and not dissimilar to the one already described.

There is another point worthy of notice. In all medical works treating of nursing, we find very few minute descriptions of physical qualities requisite for a good wet-nurse. Certain conditions are insisted upon as indispensable, such as well-developed mammary glands, strong digestive organs, good health, freedom from disease, or any particular weakness ; she must be neither too thin and spare, nor too fleshy and corpulent ; the nervous temperament is described by several writers as particularly unfavorable. We find a similarity, a correspondence in qualities, everywhere described — nowhere opposite or contradictory qualities. In fact, if we should quote the various descriptions or directions given for selecting a suitable wet-nurse, from different writers, in their own anguage, we should find that they correspond precisely with that normal standard of organization upon which we believe the law of increase is based.

The evidence derived from this source is valuable for two reasons : first, these writers have drawn those descriptions (of what constitutes a good "wet-nurse") from their own experience and observation, without any theory of their own, or any design of contributing evidence to establish a general law ; and secondly, these descriptions of what constitutes a good nurse come from a large number of medical writers of diverse character, living in different countries and writing at different periods. Such a remarkable agreement or uniformity in all their statements shows that the great facts or truths of science, wherever carefully studied and collected, not only harmonize with each other, but must have a basis or foundation in the primary laws of Nature. And further, in regard to the matter of nursing or affording natural support to the infant, it should be carefully observed, that it bears most intimate relations to other laws. As the laws of Nature come to be correctly and fully understood, we always discover a natural harmony, consistency, or adaptation to specific ends. Scarcely any truth or general principle is more firmly established than that where Nature makes a demand, she invariably furnishes a supply, and vice versa. The existence and character of the one presuppose that of the other. There may, it is true, be grades or different degrees in the matter of demand and supply; but wherever the supply is the most ample or pure, the inference or indication is clearly manifest, that it points to where the law of demand, in its best estate, has its basis and support. The natural inference then is, that the organization which is found best adapted to afford proper nutriment to the infant must be the best for its production ; or, in other words, must be regarded as the true physiological or normal standard upon which is engrafted a general law of increase. The conditions best calculated or indispensable to support life must exist necessarily in the organization that produces it. This is a universal law of Nature, supported by all experience and

observation.   Let us repeat it : the physiological conditions in Nature found necessary for furnishing the
proper nutriment for its productions, must also constitute
the same standard of organization upon which Nature, in
her normal state or highest development, has established
the ,law of production.   If, then, all the conditions or
qualifications of a good nurse in the best or highest state
are brought together, they furnish virtually the physiological or normal standard of woman for increase.
  The fourth topic for consideration is the

CHARACTER OF OFFSPRING.

While this might be considered a sequel or consequence
of the former conditions, arguments may be deduced
from this source also to establish the doctrine already
laid down.   It is scarcely necessary to state that surprising differences exist at birth in the physical qualities or
constitution of the infant ; that many are born into the
world with the seeds of disease, with weakness, imperfection, deficient vitality, organs poorly balanced, etc. etc.,
while others inherit a sound, healthy constitution, —
free, comparatively, from weakness or any natural predisposition to disease, with an organization adapted to
enjoy good health and long life.   Now what makes this
difference ?   Why are some children born with feeble and
diseased bodies, or predisposed to disease and premature
death ?   Why, in civilized society, should nearly one-third
of all infants die, the first year of their existence, and
almost one-half under five years of age ?   Is there not
something abnormal, unnatural, in such mortality of
infantile life ?   How, on the other hand, does it happen
that large numbers are born into the world with strong,
vigorous, and healthy bodies, scarcely ever subjected to disease or suffering, and live till they die from old age ?   Now,
what makes the difference in these two classes ?   Evidently

the difference in the physical stamina or constitution of the parents. Which, then, of these two classes harmonizes best with the laws of physiology in its normal state? Most clearly the latter class.

The question naturally arises, then, as far as the character of offspring is concerned, Upon what type or feature of physiology should we expect to find a general law of increase based? Would it not be upon one sound, well-balanced, and healthy in all its parts and functions, instead of one imperfect and deranged, possessing the seeds of disease and decay? Such an inference surely accords not only with all our experience and observation, but with the established laws in the orders of the lower animal economy. All the primary laws of Nature, or the fundamental principles of science, have their start from and foundation upon perfect standards. The laws that govern the human system can not be an exception to this general rule.

There is another view that may be taken from this point. When even in Nature we find derangements, imperfections, the seeds of disease, decay, and destruction, do they not clearly indicate that some laws have been violated, that there have been deviations from a more perfect standard, or, in other words, that such a state is abnormal, unnatural? As we study the present developments of human nature, we find not only a vast amount of pain and difficulty attending pregnancy and parturition, but that pain, disease, and premature decay follow their production. These, too, we find are to a great extent the common, uniform results and not exceptions to a general rule. A careful review of all the facts connected with the state and organization of infants at birth, with the nature and character of their diseases, together with the early decay and premature deaths of so many, all goes to show that if there is a general law of increase, it is certainly not based upon present standards or models. This topic — *character of offspring* — might be greatly enlarged upon, as connected with the law of increase and the designs of Nature. It has been well remarked that the two strongest instincts of man are : first, that of preserving life, and second, that of transmitting life to others. Now, if Nature has established some general law for this purpose, as she undoubtedly has, it should result in the highest development of offspring. It should pro. duce sound, healthy structures, and not an organization impregnated so much from its very origin with the seeds of disease and premature decay. It is unnecessary here to follow out the argument, that in order to perpetuate the race as it should be, there must be sound and healthy stock. There is nothing so much needed at the present day for the progress of the race, or for the advancement of civilization, as greater attention to the observance of this law.

We have now passed in review four distinct points or topics, viz : pregnancy, parturition, lactation, and offspring, which constitute the leading stages, or more prominent events, connected with the laws of propagation. The facts and inferences gathered from each of these sources all go, we believe, to show that there exists in physiology a normal standard for this purpose. Now, if by gathering up all the facts and indications to be found in each of these stages or events, we find them all in the main pointing in one direction — all agreeing with each other, and aiming at the same result — it certainly strengthens the argument, and affords an accumulation of evidence on the subject. As the four heads above mentioned seem to cover the whole ground, if not a single conflicting fact or argument can be gathered from any one of these sources — particularly when the four heads are brought together—it furnishes strong evidence in favor of a general law of propagation.

Several reflections naturally grow out of the present discussion. The subject is altogether too large and complicated to be unfolded in one short paper; all that can possibly be attempted at the present time is to present a few thoughts and suggestions upon a thesis that would require volumes for a full and thorough discussion. Inasmuch as this essay opens new views on questions which are obscure in their nature, far-reaching in extent, and upon some of which there has long existed a great variety of opinions, the sentiments here advanced should not be judged of hastily; we could wish that no preconceived opinions or prejudice should be allowed to interfere with their calm and dispassionate consideration. The only just and fair method of testing their correctness or falsity is by some definite knowledge of the subject — a knowledge obtained from the study of Nature and the deductions of facts, collected from one's own experience and observation.

The process, by which many of the leading principles of science were first established, has been slow and attended with opposition and difficulty. The more radical these principles were, and the more sweeping in their applications, the greater the contention and the strife, and the slower the growth. But whenever in the history of science any theories or principles have had a sure basis in Nature, though they might for a time be opposed and be controverted, they were sure ultimately to prevail.

So in reference to the doctrines contained in the present paper: if they constitute a part and parcel of the laws of physiology, opposition and prejudice will in time give way, and their truth and worth will come to be universally acknowledged. In fact, the history of medicine furnishes instances of new discoveries or modes of practice, which, on their first promulgation or introduction, were bitterly opposed and even ridiculed, whose truth and value came in time to be admitted, and which are now acknowledged according to their real worth and importance.

If the views presented in this paper are true, any candid person acquainted at all with the laws of physiology or the principles of medicine must admit that they are of priceless value. For illustration : In all studies, whether of Nature, science, or art, there are great advantages in having leading principles or fixed standards to guide us in our inquiries, and present beacon lights in every direction. If, while investigating the facts connected with propagation, such as the complaints of pregnancy, the difficulties of parturition, and infantile diseases, we can more clearly understand their causes and what particular laws have been violated, it must afford immense satisfaction, and might, perhaps, enable us to devise new means or agencies for relief. It will show what types of female organization are best adapted for increase, most exempt from pain and trouble, best qualified to nurse their offspring and transmit a sound, healthy stock. It will throw new light on the laws of inheritance, explaining changes which the body may have undergone in past generations, and suggesting what are some of the most fruitful sources of improvement. When we have formed a just conception of the original or normal standard of human nature, according to physiology, and keeping this standard constantly in view, — when we see the endless deviations from it, and find that these changes have taken place in accordance with the laws of inheritance,— then we begin to realize their power, value, and importance. Within a few years great interest has arisen in reference to those laws, and inquiries are being pushed in every direction for more light, more instruction, in relation to them. No one thing will infuse such interest into those inquiries and furnish so valuable a guide, as the fact that there is fixed in physiology a normal standard of propagation, from which all these laws emanate, and around which they cluster. In fact, in order to understand correctly those hereditary influences, and trace them out in all their bearings, some such chart or guide is indispensable. For, in default of some general principles to guide us, the powerful agencies of heredity can not be fully comprehended, or accurately defined, or judiciously and advantageously applied. Nor does social life, or life in any of its phases, constrain or invoke attention to any sources or agencies affecting the well-being of mankind, at least physically, which operate for good or evil, more powerfully than the laws of inherit. ance. In fact, while it is impossible to estimate the advantages of these laws when applied to human improvement, their value and application must always be limited, unless we have a perfect normal standard as a guide.

Another reflection connected with this subject is that, in attempting to account for the sufferings and difficulties attending child-bearing, and finding they arise in a great measure from changes in the human body brought about by artificial life and the violation of physical laws, the inquiries naturally arise, What are their remedies? What can be done to relieve or remove them? While we can not easily or hastily reform the present artificial state of society, or improve the physical developments of the human body, as it would require several generations to make any radical changes in this direction, yet, by understanding the true causes or sources of pains and difficulties, it may enable us to give instruction or exercise an influence that, in process of time, will tend to improve or modify these agencies, including the laws of inheritance. Inasmuch as all sanitary agencies, such as regular exercise, wholesome food, pure air, good sleep — in fact, every influence, mental and physical, which tends to improve the general health of woman — should all be encouraged as contributing to,— yea, as essential to — the realization of Nature's plan and design. The more perfect the health of woman is, the more evenly balanced her organization, the fewer weaknesses and predispositions to disease will occur, the better is she prepared for the pregnant state, for the process of labor, and the duties of maternity. All preparations or treatment that are calculated in any way to secure a normal standard of womanhood should by all means be encouraged and brought into use. There is no doubt that much has been and may be done in this way to prepare the system for these changes, and that more or less suffering, disease, and danger connected with child-bearing arise from the want of such precautions.

It should, however, be borne in mind that, inasmuch as most of these causes of pain and difficulty are the results of violated laws, extending back for generations, they can not at once be removed, and the idea that we can have "parturition without pain," as is claimed by some reformers, — especially in the present state of society — is contrary to reason. No such desirable boon can be obtained by any "course of diet" or "rules of hydropathy." It has been advocated by some, that if the pregnant woman subsisted entirely upon food free from phosphate of lime, the osseous portion of the infant — especially the skull — will become very much modified in hardness, thus making its passage through the pelvis much easier. While in some instances experiments of this kind may have proved apparently successful, in other cases they have not been attended with the same result, and as to offspring, we believe such a course of diet must prove decidedly injurious.

But attention has not alone been confined to this kind of preparation and treatment; expedients, in great numbers and variety, such as anæsthetics, medicines, instruments, etc., have been resorted to, in order to relieve the pain and difficulties of child-birth. This is well, but these are all artificial helps, relieving only for the time being; how much better is it to go further back and remove, if possible, first causes !

One of the most beneficent features, and we might say, next to life, the leading object, of the medical profession, is the relief of pain, the amelioration of human suffering. Whether, under medical treatment, disease can be cured, and life prolonged, or not — one thing is certain, pain and suffering, in all cases, can be more or less relieved.

While hitherto, in medical practice, cure has been the watchword of the profession, let, hereafter, another term, expressing a higher if not a nobler object, stand alongside of it — that is, prevention.

A gold medal was very properly awarded lately in London to the writer of an essay on "The Therapeutical Means for the Relief of Pain"; but a richer and more enduring reward, in the thanks of great multitudes, awaits the advanced guard of the medical profession, who are laboring to expound sanitary laws and diffuse a knowledge of hygiene for the prevention of pain. So in obstetric practice: while the most protracted study and greatest skill and ingenuity have for years been exhausted in devising means to relieve pain and save human life, in the most critical periods of woman's existence, let us turn our attention more to the primary causes of this suffering and danger, and earnestly inquire for preventive as well as curative treatment. Let us fully realize that, if a normal standard of physiology generally prevailed, if its principles and developments were perfectly exemplified at the present day in the human system, woman would suffer comparatively but little pain or danger at such periods.

Appendix.

[The following extracts are copied from leading medical journals in this country and in Great Britain, and are introduced here, not merely for their favorable notices of the papers referred to, but also for their independent suggestions and opinions upon the same class of subjects.]

LAW OF HUMAN INCREASE.

THE August number (1873) of the *Glasgow Medical Journal*, edited by a committee of the Glasgow and North of Scotland Medical Association, contains an extended review of Dr. Nathan Allen's pamphlets. This review quotes largely from these pamphlets, interspersed with numerous comments, from which only two paragraphs are here copied:—

"We have begun with the preceding extract from Doctor Allen's pamphlet on 'Population and its Law of Increase,' published in 1870, in order to show at a glance where the author's ideas approximate most closely to our own. This discussion of the 'general law' of increase, to which he attributes so much importance as to regard it in the light of a new discovery, presents so many points open to remark that we fear our limits will not allow of our following him in detail. But we shall endeavor to convey to our readers somewhat of the scope of a very ingenious, and as we think, very valuable, *if not in all points unassailable argument.*"

The reviewer in another place remarks that —

"The facts here brought to light are certainly curious, and may be even surprising to some of our readers. The theory adopted by Doctor Allen may or may not be a complete explanation of the facts, but it is .unquestionably one set forth upon no trivial grounds, and although we may probably think the author a little too easily satisfied as to some of his positions, it is impossible to deny him the attention due to all earnest efforts to discover truth. Furthermore, the views of Doctor Allen have an interest beyond even the present age of history."

THE RELATIVE INCREASE OF POPULATION.

ALTHOUGH the subject of the relative increase of population among the higher and lower orders of society is one of vast importance, yet it has been terribly neglected. While men are discussing the great question of the rise

22

and fall of nations, they seem to ignore the one great fact, where the men of whom nations are made are to come from.

Dr. Nathan Allen, of Lowell, is one of the few who have given the subject a scientific attention. He has studied the subject as it ought to be studied, statistically; and he has discovered that the average number of children to American families in New England is but three, or three and a half, against an average of six to eight in families of a corresponding social scale a century ago. Population can scarcely be kept up among the cultivated classes, unless an average of more than three children are born to a family. This, then, is the great problem of the future; it is a problem which no nation and which no government has yet solved.

The most pertinent fact concerning these statistics is, that they are in the main true (for Doctor Allen's opponents have made no headway against him), and they are confirmed by common observation. Practitioners in New York and vicinity tell us that they find not more than an average of three children in a family. Among the higher classes, very few families have more than five children, very many have but one, altogether too many have none. The time must come when the labors of Doctor Allen and the few interested in this matter will be appreciated, when their errors, if they have made any, will be corrected, and the whole subject will be scientifically investigated. The time must come when physicians will evince a disposition to employ a portion of their leisure in studying the great laws of population. —*The New York Medical Record, March* 1, 1872.

PHYSICAL CULTURE.

DOCTOR ALLEN's estimate of the importance to the student of an ap-propriate, well-administered physical culture, is by no means exaggerated. That there is a most intimate and necessary connection between the im-provement of the mind and the culture of the body, all will admit. It is now pretty well established by the highest authorities in medical science, that the brain is, in some sense, the organ of the mind — that all mental manifestations in this world depend very much upon the size and quality of the brain, and various agencies affecting its functions. If, therefore, all intellectual culture is not only dependent upon, but, in a great measure, con-trolled and limited by, certain physical conditions, it is of the highest importance, in the course of a liberal education, to understand what these conditions are, and to be able to turn them to the best possible account.

There are several modes or kinds of exercise in popular use, which, however otherwise allowable, are open to the objection, that they develop and strengthen mainly the extreme portions of the body. Health and strength are not synonymous terms. A person may have great strength of his limbs, or of certain muscles, but not have really good health. It is alto-gether a mistaken idea to suppose that physical culture has for its sole object the development of strength. There are other tissues and organs in the human system besides the muscular. The healthy action of the lungs and stomach is far more important than strength in the arms, legs, or back. It is in the general exercise of all the muscles, and of every portion of the body, that the system of gymnastics advocated by Doctor Allen has its great excellence. It aims to produce just that development of the human system upon which good health is permanently based, described by a distinguished writer thus: "Health is the uniform and regular performance of all the functions of the body, arising from harmonious action of all its parts — a physical condition implying that all are sound, well fitting, and well matched. Some minds do not look far enough into life to see the distinction, or to value it if seen. They fix their longing eyes upon strength — upon strength now, and seemingly care not for the power to work on; to work well; to work successfully for years to come, which is health." — *American Medical Journal, April,* 1870.

LESSONS ON POPULATION.

WE have been favored with some pamphlets by an American *confrère* (Dr. N. Allen) which are very curious. In his last pamphlet, entitled "Lessons on Population suggested by Grecian and Roman History," our author quotes Professor Seelye, who asserts that, whatever the remote and ultimate cause may have been, the immediate cause to which the fall of the Roman Empire may be traced is a physical, not moral, decay. While the aversion to marriage and the unwillingness to multiply are mentioned as becoming stronger and stronger, the historian nowhere undertakes to explain the cause of such perversity of disposition. Professor Seelye adds — "The same phenomenon had showed itself in Greece before its conquest by the Romans. Then the population had even greatly declined; and the shrewd observer Polybius explains that it was not owing to war and plague, but merely to a general repugnance to marriage and reluctance to rear large families, caused by an extravagantly high standard of comfort." "For when," says Polybius, "men gave themselves up to ease and comfort and indolence, and would neither marry nor rear children born to them, or, at least, only one or two in order to leave these rich, and to bring them up in luxury, the evil soon spread imperceptibly, but with rapid growth; for when there was only a child or two in a family for war or diseases to carry off, the inevitable consequence was that houses were left desolate, and cities by degrees became like deserted hives. And there is no need to consult the gods about the modes of deliverance from the evil, for any man would tell us that the first thing we have to do is to change our habits, or, at all events, to enact laws compelling parents to rear children."

Doctor Allen remarks that — "Some comparison may be made between the population of the United States of the present time with ancient Greece and Rome." He states that — "Whereas among the first settlers there was an average of about eight children to a family, it is doubtful whether the average number of children to each family now exceeds four. It is estimated that the number of families having no children, or only one, composes now about one-third of all New England families. Closely connected with this topic is another ominous feature of the times, that the marriage rate is relatively decreasing."

"Again," says Doctor Allen, "connected with and growing out of this selfish view of marriage, the sacredness and permanence of the institution sit lightly upon such parties. Causes for divorce are easily found. If divorces continue to multiply, as they have done for a generation past, this will certainly tend to weaken the relation and make it more and more unstable." — *The Medical Press and Circular, London and Dublin, July* 31, 1872.

PREVENTION OF DISEASE, INSANITY, AND PAUPERISM.

IN dealing with the subjects of this paper, Doctor Allen confines himself to the possible prevention. Recognizing at the outset the lack of success in agencies heretofore employed to check them, he presents, in a forcible manner, the better suggestions which from time to time have been made by the more thoughtful of those who have studied disease, insanity, and pauperism under their social aspects.

The progress in this and other countries of systematic attempts to prevent the rise and spread of disease is briefly reviewed, and the many encouraging results are noted. Doctor Allen lays great stress upon the effects of heredity, and takes the sensible view that the transmission of diseases should be prevented by care in forming matrimonial alliances. He finally quotes the words of Doctor Bowditch, namely: " Our art of the present day looks . . . . to the *prevention* as well as the cure of disease. And this is to be done by sanitary organizations throughout each state, the nation, laity, and the profession heartily joining hands in this most noble cause."

In regard to insanity, Doctor Allen urges that no organized endeavors to secure its general prevention have ever been put forth. Statistics show that only about thirty per cent. of inmates of insane asylums are cured. This is the strongest argument in favor of some attempt at prevention. The expense of sustaining the insane is augmented yearly by the increase of insanity. What can be done by way of prevention? Diffuse information on the subject among the general public; impress upon the public that intemperance, hereditary influence, ill health, fast living, high pressure in educational systems, are common causes of insanity, and set in motion means by which these causes, with the help of the public, may be modified if not overcome.

In reference to pauperism, he states that in New York 62.50 per cent., and in Massachusetts 67 per cent., is due to intemperance. The means of prevention here suggest themselves. In order to effect a decrease of pauperism, education should be enforced, communities of the vicious and idle should be broken up, surroundings should be made better, and pauper families separated and scattered. But little can be done to make a person once a pauper self-supporting. Preventive means must therefore be directed to the young who are exposed to the influences of idleness, poverty, dissipation, and filth. In this way prevention will become more helpful than any attempt at cure. Prevention, too, may be made active by directing attention especially to improvement in hereditary agencies and to the crushing out of disease.

This superficial sketch gives only a hint at much more that is interesting in Doctor Allen's paper.—*Boston Medical and Surgical Journal, July* 18, 1878.

POPULATION AND CIVILIZATION.

IN the transmission of the physical system, the organization of woman is of more importance than that of man, and thus wherever we find a high birth-rate, or a rapid increase of population, among a people, race, or nation, we are sure to find healthy women and well-balanced constitutions. The opposite of this last statement is, as we have seen, equally true. For wherever a civilized and enlightened people fail to obtain a proper increase in numbers from generation to generation, it shows something wrong and unhealthy in their domestic relations, something defective in the type of their civilization and Christianity, some violation of the great laws of health, life, and human increase. — *Journal of Public Health, London, August,* 1877.

REPORT ON LUNACY.

THROUGH the courtesy of Dr. N. Allen, Commissioner in Lunacy to the Commonwealth of Massachusetts, we have been favored with the official report for 1874. Some valuable remarks are made as to the difference between acute and chronic insanity. The causes are often complex and latent, and we are unable to determine when and how the disease originated, and consequently it is sometimes a difficult thing to state whether we have an acute or chronic malady under our observation. The question naturally arises, at what period diseases may be said to have passed from the acute to the chronic stage, and in no disease is this point more complicated than in insanity. The chances of curing the disease in a chronic form are very small. . . . . We consider the remarks of Doctor Allen on treatment extremely valuable, and we here quote them. . . . . . .

He also makes various suggestions for the management and improvement of hospitals, and we congratulate him upon his earnest labors in the field of psychology. He has given us a most valuable report.— *Journal of Psychological Medicine, Oct.* 1875. *London: L. S. Forbes Winslow, M. D., Editor.*

VITAL STATISTICS.

MANY inquiries have arisen within a few years, as to what were the causes of this decrease or stationary state of population in France, and the prevailing opinion expressed, that the causes usually alleged for such changes could not account for the present state of things. Among the various works recently published on the subject in France is one by M. Bertillon, who, after most thorough investigations, reaches conclusions similar to those of Doctor Allen, in the United States. He maintains that the French people, as a whole, are in very good circumstances, and while exercising much foresight in providing for themselves and their families, they are quite indisposed to assume greater burdens or responsibilities than need be. They have set up a higher standard of living than they have means to gratify, and they must direct all their energies and economize their means to reach as far as possible that standard. These propositions are illustrated by M. Bertillon in a great variety of ways, but the question might naturally be asked, What led to this high standard of living? What should prompt in the French this peculiar prudence and forethought in their modes of living and the objects of life? Why should they differ from their neighbors, the English and the German? May not peculiarities in physicial organization account for these changes and differences?—*The Sanitarian, New York, June,* 1877.

CONSANGUINEOUS MARRIAGES.

THIS essay of Doctor Allen's presents a very fair exposition of the present state of the question in respect to the influence exerted by the intermarriage of blood relations upon the health and integrity of the organism in the offspring of such intermarriages. Taking all the evidence bearing upon the subject that has thus far been adduced, it is very certain that the verdict must be rendered against the safety of these marriages, so far as concerns the health and well-being of the offspring.

There would seem to be some good reason for believing that the simple intermarriage of relations, continued, generation after generation, will exercise finally a deleterious influence upon the offspring, by lowering the energy of the vital forces of the system, and impairing the tone of the intellect. When a marriage takes place between two cousins, both of whom are in good health, and subject to no abnormal or morbific agencies, being the offspring of healthy parents, we assert that there is no danger that either the bodily health or the intellectual development of the immediate issue of such a marriage will be deteriorated. But when in the parties to a marriage, or in either one of them an hereditary taint or strong proclivity to disease exists, then the evil effects of the taint or the predisposition to disease they inherit will, most certainly, be entailed with augmented intensity upon their descendants by the intermarriage of the latter.

It is unnecessary to attempt to account for the ill effects of such marriages by advocating that there is some "organic vitiation" in such cases, or that there is something mysterious in the "blood of kindred." All the effects of such unions, however singular and conflicting, can be explained upon altogether more rational and satisfactory grounds. Admitting that there is a greater resemblance, likeness, similarity, in family connections, extending sometimes to almost every organ in the body, than would be in the same number of families not related, and that when these connections form matrimonial alliances, it must have, according to the laws of hereditary descent, a marked effect upon their offspring; if, in addition to this resemblance, these same parties should both have certain internal organs imperfectly developed, diseased, or predisposed to disease, the resulting evil must be still greater, the nearer the relation, and the more imperfect and diseased the bodies of both parents are, so much the more obvious and extensive will be the injuries inflicted on the offspring. On the other hand, if this resemblance in the two parties to the union is based upon a better-balanced and healthier organization, or if even one of the parties be so constituted, the favorable effects will be seen at once in the offspring. And the more perfect and healthy the organization, the better and more naturally all the organs in the system perform their functions; with other conditions favorable, the stronger, healthier, and the more numerous will be their progeny.—*American Medical Journal, Oct.* 1867.

BIRTH-RATE DEPENDENT UPON ORGANIZATION.

THERE can be no doubt that the subject of the increase of population among Americans is becoming a serious problem, when we reflect upon the manner in which the intellectual faculties of our American women are cultivated at the expense of their physical development. It is a fact which is probably accepted by all those who have reflected upon this subject, that excess of development of any part of the system is hostile to fertility. This would seem to be clearly shown by the growing disinclination among young married women to assume the duties of the maternity relation. This brings us to the paper of Doctor Allen in question, "The Normal Standard of Woman for Propagation." By the "normal standard" we understand the highest standard, or most perfect development, which physiology can present; a standard which, to be normal both in structure and function throughout, must be based upon a physical system evenly developed in every part or organ, so that each can perform its respective functions in harmony with all the rest. This standard must have its basis in the highest or most perfect development of the body as regards its anatomy and physiology. In order to understand such a law of Nature, it is shown that there are many considerations to be looked at carefully. The health and constitution of woman — the peculiar effects of gestation, and the physical changes occasioned by it — her qualifications for nursing and taking care of her offspring, and finally, the organization and character of that offspring. Primarily, we are shown that in the consideration of the whole subject we are to constantly bear in mind that child-bearing is the normal state of woman — that it harmonizes with her whole organization, and the leading features and controlling forces of her organism are evidently intended for this purpose. It is also insisted upon that the observance of this law has been found to be absolutely necessary for the most complete development and perfection of woman's organization, this having been conclusively proved by statistics on a large scale. We are shown that, if there is a normal standard in the organization of woman based upon the principles of physiology, we are to look for its effects and manifestations in pregnancy and in parturition ; in the qualifications for nursing; and finally, in the character of the offspring. Again, artificial habits may not only have reduced the vital energies of the system, but changed the size and structure of the pelvis itself, so as to interfere seriously with parturition. Doctor Allen justly complains that in a high state of civilization there is a large amount of indolence, false modes of living, injurious styles of dress, and other evil practices that interfere, not only with a natural and healthy state of the whole body, but concentrate their evil effects particularly upon the pelvic organs, and that these effects, by inheritance, become greatly intensified in their form and extent in successive generations. We would commend Doctor Allen's paper to the careful and thoughtful consideration of every physician who takes an interest in this important subject, and think it will amply repay careful perusal.— *New York Medical Journal, September,* 1876.

SANITARY SCIENCE.

In an address delivered on the 5th of October, 1887, at Toronto, Canada, before the American Health Association, by Dr. Nathan Allen, one of the most prominent views of this question is discussed.

" It is just to state here that the claims of the sick poor have been most liberally responded to by physicians, and that no other profession or class of men do so much for the poor as the medical profession. But this work of charity has its equivalents ; — it secures to the physician a stronger hold on the affection and confidence of the people, and, in different ways, tends to increase his business. But to engage actively in means to *prevent disease,* not simply in one instance, but in case of great numbers, this is very different,—it cuts off directly the support of the physician."

There has never been any inscrutable mystery about the treatment of disease. Charlatans have tried to cloak their ignorance with pretense, but the true professors of the healing art have in all ages and countries rather erred on the side of loquacity as to their views and principles of procedure, than hidden or disguised them. Whilst it was believed that disease was something like an entity to combat and struggle against, using drugs and the appliances of art as weapons, this theory found ample expression in the works and utterances of the doctors. Now that it is felt that the re-establishment of health is the true method of procedure, that view is expounded with equal liberality; and as it goes without saying that if what will prevent or correct departures from the line of health-preservation must be better than cure, so by a perfectly natural process, without any exceptional philanthropy or shortsightedness on the part of the medical profession, *physic comes to be matter of sanitation.* This we take to be the true explanation of "The Relation between Sanitary Science and the Medical Profession," which formed the subject of Dr. Nathan Allen's address at the Health Association. — *London Lancet, Dec.* 1887.

MARRIAGE RELATION.

IMPELLED by a sense of duty, I venture to touch upon a matter of extreme delicacy, but of vital interest. It is asserted by an American writer (Doctor Allen) that in *certain* classes of society in some parts of the adjoining Union, for a long time past, the marriage relation would seem to be regarded not as a Divine institution, ordained by God for the preservation of the species, but as a matter of convenience and self-interest. To use his own words: " The standard of living is too high ; the artificial wants are too many ; confinement to household duties is irksome ; children are a burden ; the responsibilities of maternity must be avoided or limited. Hence, in married life, a series of nameless acts takes place, which need not be described." In those few, grave, weighty, momentous sentences, gentlemen, is a picture of some of the chief causes of that alarming decline of birth-rate, and with it, and as a consequence of it, a gradual and pernicious change in the female organization. This, in thoughtful minds, has created alarm, lest the *induced* organization become permanent in type. — *Transactions of the Canada Medical Association, for* 1877. *Annual Address by its President, Dr. Wm. H. Hingston.*

WANTED: A WET-NURSE.

WE recollect in bygone days having heard a serio-comic song beginning —

" Wanted: a governess, able to fill
The post of tuition with competent skill."

338

APPENDIX.

And the song goes on to reckon up so many kinds of knowledge, useful and ornamental, such as good health and niceness of appearance, and such good temper and patience, amongst the qualities desired, that the young man who sings the song declares at the end, that when he finds so perfect a creature he shall make her his wife.

This song is vividly brought to our recollection as we read the papers of Dr. Nathan Allen, of Lowell, Mass. Some time ago, in an article on "National Degeneracy," we called attention to Doctor Allen's well-timed warning against the causes which war against the health and fertility of his own countrywomen — warnings equally useful on this side of the Atlantic. In a paper lately published he enlarges on the same topic, and amongst other points, dwells so strongly on the characteristics of a good wet-nurse that the evident corollary is, that men who marry would do well to seek for these characteristics in their wives. " No fact in vital statistics," he justly tells us, "is more firmly established than that, in order to promote health and save life, the infant should be nursed at the mother's breast." Chemists and physicians have tried in vain; nothing can be found which is a substitute for " pure breast milk." For this Nature has made ample provision in the normal organization of woman. In a perfect woman, " the lymphatic and sanguine temperaments " are well developed, "together with vigorous and healthy digestive organs." "A good nurse must have well-developed mammary glands, strong digestive organs, and even, uniform disposition. Such must have good health, freedom from all disease, no particular weakness nor excessive nervousness." Qualities, these, not to be despised! — and we must say that, although a wet-nurse may not have all the qualities of a wife, she should have all those of a wet-nurse.

Unfortunately, Doctor Allen is obliged to confess that many of his countrywomen can not be good nurses, because they have " neither the organs, nor nourishment requisite for even a beginning." He looks upon this as but a symptom and a part of that general want of the power of producing and rearing a lively offspring, which he says marks the American woman side by side with her more favored Teutonic and Celtic sisters, and which threatens to put in a minority the native Americans, the descendants of the early, hardy, and prolific settlers.

As the outcome of his observations, Doctor Allen propounds a law of human increase to the following effect — in order that a people shall increase, it is necessary that their organisms shall be well and harmoniously balanced. Excess of development of any part of the system is hostile to fertility. It is not sufficient that the population be not sickly, but one set of organs must not preponderate too much. "Families in which are found genius, great talents, and supreme devotion to intellectual pursuits are not prolific." The families of "great scholars, authors, poets, or nearly all who have been eminent or distinguished in any department of science," generally become extinct.

In the case of the New England women, according to Doctor Allen, a double force acts against fertility. "If," he says, "that portion of the brain whose functions include attachment to the other sex, love of offspring and domestic life — those strong instincts which center in the family and the home — is not properly developed and trained, but other portions of the brain, embracing the selfish faculties, are continuously exercised," the whole womanly organization is changed. "It tends to undermine the foundation of the marriage and maternal relations, which rest on the purest and most powerful instincts of Nature, and transfer it to one of interest and convenience. The relation is, in fact, already coming to be viewed more in the light of a partnership, as a matter of business and necessity; or, in other words, to be based upon the supremely selfish elements of human nature. That such large numbers of our young married women should be disinclined to assume the duties of maternity indicates something wrong. However desirable or important may be the cultivation of the mental faculties, these alone never bind and cement society permanently in the home and in the family. In such cases children are a burden, home is irksome, domestic work not agreeable."

Thus Doctor Allen alleges that in the training of some of his country-women the faculties conducive to fertility are repressed, whilst the intellectual faculties are cultivated to excess; and these, as Prof. Samuel Haughton long ago demonstrated, use up far more nourishment than muscular work does. The women are brilliant and spirituelle, but lacking in bone and muscle. They do not deserve the Homeric epithet, "deep-bosomed." We may observe that Doctor Allen equally denounces stupidity and mere animalism as contravening the law of harmonious development.

It is very significant that such warnings on the education of women in relation to the increase of population should come to us from America. It is a warning which we may well accept; and young men, when they think of marriage, may bear in mind Doctor Allen's law of the need of even development of the whole organism; and that, although no man should ally himself with an animal or a fool, yet there comes a time in every happy marriage when the brain may rest awhile, and when "wanted: a wet-nurse," will be the word unless the wife has her share of the most beautiful and bountiful part of the female organism. — *London Medical Times and Gazette*, 1886.

PHYSICAL DEGENERACY.

AMONGST other elements, whether causes, effects, or evidence, of degeneracy, Doctor Allen notices the inordinate passion for riches; overwork of body and mind in the pursuit; undue hurry and excitement in all the affairs of life; intemperance in eating and drinking; the enormous use of quack medicines; the general indifference to human life; the increased use of spirits, tobacco, and opium; the increase of lunacy; the decrease of children.

Fashionable dress obstructs the play of the lungs and displaces the pelvic viscera. Connected with this weak and relaxed state of the muscular tissues, and the above-mentioned effects of fashion in dress, has sprung up a class of very grave complaints, which once were comparatively unknown in our country, and are somewhat peculiar to American women. We refer particularly to weaknesses, displacements, and diseases of organs located in the pelvis. Within twenty or thirty years, this class, comparatively new, has increased wonderfully. No one but a medical man who has devoted special attention to this subject, can realize fully what are the nature and extent of this change, and what are the direful effects. These same complaints have frequently been produced — have certainly been aggravated, and sometimes made incalculably worse — by the various means and expedients which the parties have resorted to, to interfere with or thwart the laws of population.

From physical to moral degeneracy the *descensus* is but too *facilis*. Besides the inherent defects in such an organization, in not making the necessary provisions for gestation and lactation, the natural instinct of woman in a pure love of offspring and domestic life become changed; the care and trouble of children are a burden; society, books, fashion, and excitement generally are far more attractive. The results will be — first, sterility, for no intensely nervous temperaments are favorable to increase; and, secondly, an increasing ratio of degenerate population. It is woman that "moulds the physical systems of those who are to come after us," and "imperfect developments are transmitted in an intensified form." Meanwhile, the mortality of women, instead of being lower in America than that of men, as is the case in the Old World, is higher.

We have thus far followed Doctor Allen, not with any view of giving a disparaging account of American life, but of pointing out that, if like causes produce like effects, we had need take heed ourselves. There is not one of the evils denounced by Doctor Allen which we do not find here. Large cities are the graves of our population. Women can not be developed, nor children be reared, in them. No one who looks at the stunted race of London "roughs" and "costermongers" can feel that we have any thing to boast of. Even our peasantry are not models.— *London Medical Times and Gazette, March,* 1871.

PHYSICAL ORGANIZATION — ITS EFFECTS.

AT so many points do the observations of Dr. Nathan Allen touch this subject, that this appears the proper stage for their consideration. They are worthy of investigation upon other grounds. They embrace what the statisticians have unwittingly omitted or purposely ignored; they describe many of the causes and circumstances which must, in a greater degree than mere relationship, influence fecundity, sterility, and the production of healthy offspring, without which, in fact, no exhaustive or comprehensive estimate of degeneracy can be obtained; and they supply a practical illustration of the speculations as to the perpetuation of races from an unexpected quarter. The attention of the author has been directed to this inquiry by the rapid decrease in the native inhabitants of New England, which, should it proceed at the same rate for another hundred years, will in all probability extinguish the descendants of the sturdy and stalwart pilgrim fathers.

That a different rate of propagation is characteristic of different races, or the same race living under different circumstances, has been demonstrated. That luxury, licentiousness, intemperance, and the antagonistic conditions of frugality, virtue, and asceticism, exercise, or may exercise, an important influence over the rate of fecundity, has been admitted; but in all such investigations the state of the organization and of the functions of the parents have been singularly overlooked, and information sought for from remote or collateral sources. The author conceives, and rightly conceives, that the health of the progenitors determines, more than any other circumstance, the numbers, the vitality, and the state and strength of the organs and all the functions, mental and physical, of the issue.

It is perfectly clear that mature and undiseased stock is required for the production of healthy descendants. In this sense the effect of hereditary tendencies may be admitted, as, while we may refuse to believe in the transmission of a specific taint, it is obvious that the feeble, the exhausted, the consumptive, the syphilitic, must be less capable of procreating at all, and of producing strong and uninfected progeny, than the virile and vigorous. In a similar acceptation, propinquity may be held to act by intensifying whatever may be most prominent or potent for good or for evil in the constitution of the descendants, and barrenness may thus depend, not upon impotence created by kinship, but by weakness and disease.

It would appear that states of the system much less important than positive or structural diseases, whether innate or communicated, affect or limit reproduction, as, for instance, changes in the balance of nutrition, as in corpulency — the relative activity of different organs, as where intelligence is highly cultivated by intense study, where the imagination and emotion are excited by literature and refinement at the expense of the muscular and digestive power, and where all the appliances of modern civilization are employed in stimulating and thus exhausting the nervous system. As a corollary to this proposition, it may be mentioned that giants and dwarfs are unprolific, and that persons of expanded or contracted mind have few descendants. Excess in the gratification of the propensities is visited by a similar retribution, and the vicious, the dissolute, the intemperate, are generally the last of their line. — *The British and Foreign Medico-chirurgical Review*, 1875.

William A. Stearns, D. D., LL. D.

[SOME men live in advance of their times. Such was William A. Stearns, to whom this book is dedicated. The early establishment of physical culture in Amherst college was much indebted to his efforts while president of the college. In evidence of his deep interest in physical education and his desire of introducing into the institution measures of some kind for the better training and developing of the human body, we make the following extract from his inaugural address, given Nov. 22, 1854 : —]

Education, therefore, may be contemplated in the first place physically. It involves the developing and energizing, at least the protection, of the physical system.

We can not expect that all men will be like the first pair in paradise,

> — of noble shape, erect and tall,
> Godlike erect, with native honor clad.

But we can expect a much higher measure of physical perfection than is ordinarily attained. Much depends upon it, duty demands attention to it.

Bodily disarrangement is not only occasion of suffering, but often of moral perversity and intellectual inferiority. It clouds and clogs the understanding, sometimes dethrones the reason. When the mind is not wrecked, it is enfeebled by it. Great undertakings are prevented, and ordinary affairs inadequately performed. Bodily disorder perverts the judgment. We can not justly weigh and balance considerations under the influence of it, and form safe conclusions. It is a prolific source of moral evil. It induces restlessness, stimulates bad passions, and prompts to vicious indulgences. A morbid appetite for intoxicating drinks and for hurtful narcotics is often occasioned by it. From the same source spring much envy, spleen, and misanthropy. He who intelligently offers the prayer, "Lead us not into temptation," will pay attention to his bodily condition ; for it requires less effort to be a good man, with a sound body, than with a system imperfectly organized or disordered.

Good taste teaches the same doctrine. We admire most that which approaches nearest its own perfection. This is true in horticulture, in agriculture, in ornithology, and in the treatment of domestic animals. But many a man who would spend hours every day in tending and grooming a favorite racer will abandon his children, except in actual sickness, to almost total neglect.

Anciently it was not so. The palæstra, the gymnasium, the chase, the exercise of the camp, though intended for the increase of military efficiency, promoted physical strength. The ancient ideals of perfect commonwealths have given prominence to the subject of corporeal vigor, in their systems of education. In the middle ages, too, hunting, war, the spirit of chivalry, secured both among the nobility and the masses a superior physical development. But in our country there is reason to fear that, in this respect, we are deteriorating. Partial deformity, the languid step, stooping shoulders, cadaverous countenances, are too common. Among students, has not death held his terrible revels in our day, to an extent never before realized ? Our halls of justice, and still more our pulpits, are thronged with invalids.

Physical education is not the leading business of college life, though were I able, like Alfred or Charlemagne, to plan an educational system anew, I would seriously consider the expediency of introducing regular drills in gymnastic and calisthenic exercises. If agricultural and mechanical operations, and even martial movements, could be added without injury to scholarship, so much the better. At all events, I would take measures for imparting hardihood and the proper use of the muscular energies. But without innovation, something can be done in this direction. The general laws of health can be imparted, and some of them insisted on. It can be shown to the scholar that it is not often intellectual exertion, even though intense, that digs the premature graves of students. It is neglect. It is imprudence. It is irregularities. It is sinful indulgences. It is violence, perhaps in many cases innocently committed, against the laws of the constitution.

Perhaps I am dwelling too long on this topic. But of one thing I am certain : the highest intellectual efficiency can never be reached, the noblest characters will never be formed, till a greater soundness of physical constitution is attained.

Index.